Gulbahor Abdrimova

Creation and introduction of mulberry silkworm hybrids

Gulbahor Abdrimova

Creation and introduction of mulberry silkworm hybrids

(In conditions of Karakalpakstan)

ScienciaScripts

Imprint

Any brand names and product names mentioned in this book are subject to trademark, brand or patent protection and are trademarks or registered trademarks of their respective holders. The use of brand names, product names, common names, trade names, product descriptions etc. even without a particular marking in this work is in no way to be construed to mean that such names may be regarded as unrestricted in respect of trademark and brand protection legislation and could thus be used by anyone.

Cover image: www.ingimage.com

This book is a translation from the original published under ISBN 978-620-7-46541-5.

Publisher:
Sciencia Scripts
is a trademark of
Dodo Books Indian Ocean Ltd. and OmniScriptum S.R.L publishing group

120 High Road, East Finchley, London, N2 9ED, United Kingdom
Str. Armeneasca 28/1, office 1, Chisinau MD-2012, Republic of Moldova, Europe
Printed at: see last page
ISBN: 978-620-7-76824-0

Contents

In this monograph is written about breeding of mulberry silkworm in areas with extreme environmental conditions and global changes in weather and climate conditions on the globe, urgently dictate the need for intensive development of new methods of selection and reproduction of plants and animals adapted to the changed environmental conditions. Creation of breeding populations providing normal level of growth, development, reproduction, silk productivity in less favourable conditions reflects the interests of silk breeding industry in obtaining high cocoon yields in regions differing significantly in zonal features.

The monograph is intended for undergraduates, postgraduates, senior researchers and science educators.

This monograph was written based on the results of scientific research on "Creation and introduction of mulberry silkworm hybrids for extreme environmental conditions".

Reviewers:

U.Daniyarov - Doctor of Agricultural Sciences, Professor

B.Zholybekov - Doctor of Agricultural Sciences, Professor

Recommended for publication by the Academic Council of the Karakalpak Institute of Agriculture and Agrotechnology (Minutes No. 3 of 6 February 2024)

INTRODUCTION

According to the International Silkworm Commission, Uzbekistan is the third largest producer of cocoons in the world (1.2% of world production), with a yield of 56.9 kg per box of gren[1] . The geographical location of Uzbekistan makes it attractive and, at the same time, problematic for mulberry silkworm breeding, as Uzbekistan has zones with favourable and difficult environmental conditions.

Breeding of mulberry silkworm in areas with extreme environmental conditions and global changes in weather and climatic conditions on the globe, urgently dictate the need for intensive development of new methods of selection and reproduction of plants and animals adapted to the changed environmental conditions.

Undoubtedly, the Republic of Karakalpakistan differs significantly not only in weather and climate, but also in soil characteristics, water availability and other conditions from other regions of the country.

Creation of breeding populations providing normal level of growth, development, reproduction, silk productivity in less favourable conditions reflects the interests of silk breeding industry in obtaining high cocoon yields in regions differing significantly in zonal features.

Thus, creation and introduction of mulberry silkworm hybrids for extreme ecological conditions is an urgent problem of silk breeding.

The demand for the performance of research work within the framework of the dissertation arises from the specific tasks set out in Presidential Decree PP-2856 "On Measures to

[1]www. Inserco.org/.

Organisation of the activities of the Association of Uzbekcipaksanoat" [1;p.1-5] [1;p.1-5] dated 29.03.2017 and PP-2460 "On measures for the development and reform of agriculture in 2016-2010" [2;p.5-7] dated 29.12.2015 and in the Decree of the Cabinet of Ministers of RUz No.616 dated 11.08.2017. "On the programme of measures for comprehensive development of silk breeding industry in 2017-2020" [3; p. 5-6] [3; p. 5-6], to increase the production of grena and cocoons of domestic breeds and hybrids, to improve their quality corresponding to international standards and, thus, to increase the export opportunities of the silk industry.

The Strategy of actions for further development of the Republic of Uzbekistan for 2017-2021 provides for the development of agriculture, including silkworm breeding. In this aspect, genetic breeding research in the direction of development of new highly effective methods of selection, breeding work, as well as breeding of new breeds and hybrids of mulberry silkworm with high productivity and quality properties of cocoons in extreme environmental conditions is of particular importance.

A number of governmental decrees define the tasks of silk breeding and measures to address them. For example, the Decree of the President of the Republic of Uzbekistan "On measures to organise the activities of the association "Uzbekipaksanoat" dated 29 March 2017 states that one of the priority tasks is "Creation and introduction of highly productive breeds of mulberry silkworm genetics, step-by-step increase in its production to reduce imports and fully meet the domestic needs of the Republic in the future". This work contributes to the solution of this task and is carried out in accordance with the priority directions of development of silkworm breeding in Uzbekistan.

There is no doubt that mulberry silkworm breeds of Uzbek selection during tens and even hundreds of generations of breeding have formed genotypes and populations that have retained increased survival rate, effective conversion of specific feed into silk in less favourable conditions of high temperature and low relative humidity in our region. However, despite the fact that Uzbekistan has accumulated a great experience in creating breeds and hybrids of mulberry silkworm, combining increased yield and cocoon quality in temperate breeding zones, [32;p.21-25], [38;p.102-111], [52;p.3-19], [58;p.3-100], [76;p.4-66], [85;p.3], [89;p.1-49], [93;p.42- 51], [97;p.59], [104;p.143], [108;p.3-7], [113;p.28-31], [114;p.3-20], [121;p.92], [124;p.31], [129;p.51-54] until now there were no breeds and hybrids of mulberry silkworm recommended for foraging in extreme ecological conditions of Karakalpakistan.

MAIN PART

CHAPTER I

LITERATURE REVIEW

The geographical peculiarity of Uzbekistan is the presence of different ecological and climatic zones on the territory of the country, differing in air temperature, presence or absence of water resources, and different degrees of solar insolation. This dictates the need to create special varieties of plants and breeds of animals oriented to a particular terrain. Only in this case it will be possible to obtain sufficient yields of agricultural products, including cocoons of mulberry silkworms.

At present, the majority of zoned hybrids of mulberry silkworm correspond to temperate climatic zones: Fergana, Andijan, Tashkent, Namangan, Samarkand, Jizzak, Syrdarya regions. For ecologically unfavourable regions such as Karakalpakstan, Surkhandarya and Kashkadarya regions there are no silkworm hybrids and silkworm varieties adapted to their climatic conditions. Therefore, cocoon yield in Karakalpakstan (56.8 kg), especially in some districts (Chimbay-41.7 kg), is much inferior to the cocoon yield in the whole country - 57.8 kg.

The aim of this study is to select breeds and hybrids of mulberry silkworm, which are able to fully realise their genetic potential in difficult environmental conditions of Karakalpakistan. The final goal of the work is to carry out a large-scale breed change on the territory of the republic with the introduction of mulberry silkworm breeds and hybrids recommended in this study into production.

The set objectives imply the analysis of economically valuable properties of a huge number of breeds and hybrids, created in different years in the Research Institute of silk breeding. Today the Institute of silk breeding is the most important scientific centre in Central Asia, where theoretical bases of selection, genetics, breeding, breeding, grenin production, disease control, mechanisation of silk breeding are developed.

A collection of mulberry silkworm breeds has been established and maintained at the Scientific Research Institute of Silkworm Breeding (SRI) since its organisation (1927). Among silk-producing regions of Central Asia, this is the only living collection, to which breeds from different geographical zones - Japan, China, Azerbaijan, Georgia, Ukraine, Russia, etc. - have been imported for many years. The living collection of mulberry silkworm breeds is a unique collection of the world gene pool of this important scientific and agricultural object. The collection contains 120 breeds from 12 ecological zones of the world. It represents almost the entire genetic diversity of the mulberry silkworm.

Each of these groups is characterised by a certain duration of life cycle, constitution of individuals, resistance to diseases, yield, size, shape and quality of cocoons, etc. An important characteristic of a breed is its voltinity (voltinism), i.e. its ability to produce one (monovoltinism), two (bivoltinism) or more (polyvoltinism) generations per year under conditions imitating natural ones (lower temperature in spring incubation and higher temperature in summer incubation). In most breeds, the larval stage is divided by four moults into five instars, but there are breeds with three moults.

The collection of mulberry silkworm breeds at NIIH is a scientific base for setting and solving many fundamental and practical problems of silk breeding. The interest to the world collection of mulberry silkworm NIIH has always been manifested. Scientists have repeatedly addressed to the materials of the unique collection of genetic fund of this important scientific object. For example, components of hybrids Ipakchi 1 x Ipakchi 2, Ipakchi 2 x Ipakchi 1, widely introduced in production, were created on the basis of collection breed SANIISH 30 with attraction of parthenoclones as improvers. Components of zoned hybrids AGU-112 x UzNIISH 9, UzNIISH 9 x AGU-112 intended for repeated fattening were created from the collection breeds TashSKHI-112 and SANIISH-9 [52;p.3-19]. All currently available parthenogenetic clones were obtained from collection breeds SANIISh 30, Mechnaya 1, Mechnaya 2, Baghdadskaya [110;p.3-324], [19;p.136-140], [113;p.28-31], [50;p.3-27], [49;p.25-27]. All currently available sex-labelled breeds were created on the basis of collector breeds of SANIISH 30, Biv.111 and others and genetic lines [106;p.52-72], [107;p.5-6], [135;p.9], [132;p.3-42], [112;p.343-349], [113;p.28-31], [108;p.3-7], [103;p.50].

Parthenogenetic clones and sex-labelled breeds are used to create new breeds and new highly heterotic, 100% pure hybrids [24;p.130-141], [46;p.22-24], [103;p.50], [105;p.145], [75;p.35]. The breeds of the collection possessing fine cocoon thread are used in the creation of new breeds and hybrids of mulberry silkworm [129;p.51-54], [72;p.45-51]. Components of Tetrahybrid 3 Belokokonnaya 1, Belokokonnaya 2, SANIISH 8, SANIISH 9, widely zoned in due time, have their origin from collection breeds Chinese 108, Yaponskaya 127, Bagdadskaya [97;p.59].

For successful solution of the tasks set in the study, first of all, it is necessary to introduce fundamentally new for Karakalpakstan highly viable and yielding lines, breeds and hybrids of mulberry silkworm. Therefore, special attention was paid to genetically modified breeds and parthenogenetic clones, never bred in Karakalpakstan before.

§ 1.1. Use of sex-labelled breeds and
parthenogenetic clones of mulberry silkworm in hybridisation

It is well known that the mulberry silkworm is bred worldwide only as first generation hybrids in order to maximise the heterosis effect [153;p.9-23], [154;p.29-38], [155;p.11-17], [160;p.56-62], [162;p.307-310].

Unfortunately, serious obstacles stand in the way of maximising the effect of mulberry silkworm heterosis, in particular, the impossibility of obtaining pure hybrids uncontaminated by parental material [158;p.3915-3923], [161;p.187-192], [163;p.42-59].

The preparation of hybrid worms for industrial fattening, not contaminated with purebred eggs, is a technically difficult task. The fact is that mulberry silkworm moths mate immediately after emerging from their cocoons and thus produce purebred grens. To avoid mating within each breed, it is necessary to separate females and males from each other in advance, even before they emerge from the cocoons, in order to cross females of one breed with males of another. Tens of millions of individuals are subject to sex separation. Meanwhile, the methods of sex separation of breeding material for the purpose of hybridisation are either inaccurate or very labour-intensive.

For example, the method of division of cocoons by sex used in Uzbekistan, based on weight differences between opposite sexes, due to overlapping weights of females and males, allows less than half of the separated females and males to be allocated to a tribe from a batch of cocoons, and then with a large error in the group of each sex.

Analyses of industrial gren showed that it contains only 20-25% hybrid eggs, while the remaining eggs are maternal source breed [110;p.3-324].

The clogging of the hybrid grena with the original parental breeds leads to a decrease in cocoon yield, increases the heterogeneity of cocoon raw material, disorganises the work of breeders aimed at breeding breeds that give, first of all, highly productive hybrids. Finally, silk breeding is deprived of the prospect of using breeds and lines that give powerful heterosis in hybrids. It is theoretically and practically proved that in all agricultural objects especially highly heterosis hybrids are obtained as a result of crossing somewhat depressed forms. It is quite clear that contamination by weakened parental forms of industrial hybrid grena by 75-80% so reduces the yield that further introduction becomes impossible. However, the development and improvement of silk breeding is unthinkable without accurate preparation of industrial hybrid grena. This is achievable only with precise separation of elite material into separate groups of females and males. This problem can be solved by genetic methods.

The development and improvement of silk breeding is unthinkable without the precise preparation of industrial hybrid grens. This is achievable only with accurate separation of elite material into separate groups of females and males. For this purpose, research on breeding by the radiation method a breed labelled by the sex characteristic of egg colouring was carried out at NIISH. These studies were carried out by [106;p.52-72]. [107;p.5-6], [77;16], [78;p.12]. As a result, sex-labelled breeds have been obtained in which the female eggs have a normal dark colour, while in the male eggs no pigment is produced and they appear straw yellow. Because of this, sex is unmistakably recognised on the 2nd day after egg laying by the butterfly, as soon as pigment is formed in the cells of the serous membrane. Eggs can be automatically separated by sex by high-performance instruments using an "electronic eye".

Examples of successful application of such breeds are hybrids C-13 x C-14, C-14 x C-13, zoned in some regions of Uzbekistan with the

1989 (invention certificates No. 9003649 and No. 9003630), as well as hybrids Meechenna 1 x Meechenna 2 and Meechenna 2 x Meechenna 1, introduced into production since 1990 (invention certificates No. 9103805 and No. 9103813). Breeds Meechennaya 1 and Meechennaya 2 are divided by sex based on the colouration of caterpillars.

The presence in the mulberry silkworm collection of ready-made, sex-determined breeds provides conditions for the creation of hybrids with 100% purity of grena preparation. Many of these breeds were reproduced in the collection under the group breeding scheme, therefore, reductive selection was carried out with them.

The use of sex-labelled breeds to create hybrids with 100% purity of grena preparation is only one of the many options that can be obtained by using collection breeds. For example, good effects are observed when using sex-labelled breeds and parthenogenetic clones as components to create 100% pure hybrids.

Proposals to use female clones as one of the partners of industrial hybridisation arose immediately after the development of [19;pp.136-140],[22;pp.77-81], the ameiotic method of

parthenogenesis, producing only female offspring. And this is not accidental. The point is that the use of parthenoclon as one of hybridisation partners excludes completely breeding work with this material, because the genotype of the clone does not change in successive generations, there is no need for sex separation and the reproduction rate increases approximately twice due to the fact that the parthenoclon grena develops into a female only. Such maternal forms are especially valuable for crossing with males of special genetic breeds that carry lethals on sexual Z-chromosomes and give hybrids of male form when crossing with females of any breeds. In this case it is not necessary to prepare twice as much breeding material at the previous stages, as it is done at breeding of bisexual breeds. This is indicated in their works [166;p.49], [167;p.53], [159;p.3- 50].

It is known that clones lag behind the normal breeds in productivity due to depression caused by parthenogenetic reproduction, which is not typical for this species. However, the world collection of mulberry silkworms contains parthenoclones with normal productive qualities for females [76;p.4-66].

The technical difficulties of activating many hundreds of thousands of clutches for conversion to the parthenogenetic pathway have now been resolved.

All this indicates real prerequisites for the creation and application of parthenogenetic ameiotic clones for industrial hybridisation of mulberry silkworm [103;p.50], [113;p.28-31], [110;p.144-170] Previously, a clonal-breeding hybrid 5140pk x C-5 was already created, which since 1992 was introduced in some regions of Uzbekistan and gave a good yield.

In addition, one should keep in mind the advantages of hybridisation with clones. Parthenoclones are represented by only one sex - female, so there is no need to divide cocoons by sex. The breeding plants receive cocoons from which only females are known to emerge. The second component for hybrids can be males of almost any breed. Since males fly out of cocoons earlier than females, it is not very difficult to collect them. Therefore, there is no need to carry out the expensive, time-consuming and very inaccurate operation of separating the cocoons by sex. Thus, all conditions for the creation of 100% pure hybrids, uncontaminated by parent breeds, can be easily organised at the breeding farms.

The use of part-breed hybrids in the silk industry has several very important advantages.

1. FI hybrids obtained from crossing parthenoclone females with conventional males differ from conventional hybrids not only by greater hardiness, but also by equalisation in all traits. This is due to the fact that all the females involved in the cross are genetically identical copies. The genotype of the parent breed does not change over generations, so industrial hybrids must remain constant from year to year in terms of both conventional economic-value traits and heterosis.

2. Genetic constancy of parthenoclon females allows to completely exclude labour-intensive expensive breeding work in research institutions and at breeding silkworm breeding stations. The cycle of this work is equal to 4 years. At the same time, the constancy of females also makes it possible to sharply increase the relative volume of cocoons selected for breeding from the total mass of all cocoons received, as selection among them loses all sense. Meanwhile, with the usual technology of grena preparation at the super-elite level, about 15% and at the elite level - 40% of cocoons are taken for breeding from the total volume of the received breeding batch of cocoons, each kilogram of which is much more expensive than a kilogram of industrial cocoons.

3. The presence of single females in parthenoclones is also useful for production, as it allows the preparation of hybrid grens of 100% purity without resorting to the usual very time-consuming, expensive and very inaccurate methods of separating females from males at the cocoon stage for the purpose of interbreed hybridisation. For crossing with parthenogenetic females, males of a breed with sex-labelled grens can be used. They can be taken in half the number of males for the purpose of double utilisation.

4. The propensity of individuals to parthenogenesis correlates with high combinatorial ability in hybrids, which enhances the heterosis effect.

Besides, parthenoclones can be used in breeding work. Among hybrid cocoons the best ones can be selected, virgin females of which can be stimulated to parthenogenetic development. The hybrids thus cloned will be crossed with males of appropriate breeds. The outbred offspring of such families will then be used for further selection, as is usual with the individual selection method. Numerous experimental data testify to good inheritance of silkiness of parthenogenetic mothers to both sexes offspring obtained from crossing clones with ordinary males. Therefore, already 4-5 generation of such lines will be sufficiently consolidated and will serve as a basis for the creation of a new breed, At the usual method of selection of parental pairs (family breeding, long-term rigid selection, testing of the hybrid) to create a hybrid takes 8-10 years. The advantages of using parthenoclones in breeding are obvious.

The economic effect from introduction of pure clonal-breed hybrids can be high, because nowadays the clogging of hybrids with original breeds reaches, according to different sources, 50-70% [110;p.3-324], i.e. the advantages of hybridisation in silk breeding are used not more than 30%. The use of clonal-breed hybrids will lead to 100% heterosis.

The economic effect from introduction of pure clonal-breed hybrids can be high, because nowadays the clogging of hybrids with original breeds reaches, according to different sources, 50-70% [110;p.3-324], i.e. the advantages of hybridisation in silk breeding are used not more than 30%. The use of clonal-breed hybrids will lead to 100% heterosis.

The advantages of part-breed hybridisation have already been discussed. Now it should be said that participation of sex-labelled breeds in hybrids makes the task even easier, because the breeding plants receive parthenoclon cocoons consisting only of females and cocoons of males of sex-labelled breeds. When sorting clonal cocoons for breeding, the majority of cocoons (83-87%) remain for breeding, as these cocoons are genetically homogeneous. The cocoons of males of the sex-labelled breed are sorted according to the existing rules. Hybrids between the clone and the sex-labelled breed are easy to prepare and have 100% purity [113;p.28-31]. The world collection of mulberry silkworm breeds can provide material for creating 100% pure hybrids [76;p.4-66].

All these facts are of special interest and indicate a real possibility to use sex-determined breeds and parthenogenetic clones for industrial fattening in Karakalpakistan.

Silkworm breeders have created many breeds of mulberry silkworm, which differ from each other by certain features [85;p.3], [86;p.3], [105;p.145], [102;p.79]. Thus, in sex-labelled breeds at the grena stage, female caterpillars emerge from the dark grena and male caterpillars emerge from the light grena. In the breeds labelled by sex at the caterpillar stage, females are with masks, and males are of white-milk colour, without masks. Some of the created breeds are of industrial importance, many breeds are preserved in the living collection of mulberry silkworm and are used in breeding - genetic and research works.

Silkworm breeds are divided into three groups according to the type of development: polyvoltine, bivoltine and monovoltine. Polyvoltine breeds produce three to eight generations per year, they are very viable but low yielding. Bivoltine breeds give two generations per year, they are more viable compared to polyvoltine breeds and are higher yielding. Monovoltine breeds give one generation per year, they are less viable than polyvoltine and bivoltine breeds, but they are higher yielding, characterised by increased weight and silkiness of cocoons. However, they are also characterised by higher requirements to keeping conditions.

Researchers, when selecting material to create new breeds and hybrids for spring and re-breeding, always consider extreme conditions Central Asian republics and economic and biological traits of this or that breed. For example, monovolt breeds SANIISH-8 and SANIISH-9 [124;p.31] were used in production for spring fattening and were components of complex hybrids - Tetrahybrid-3 $ (SANIISH-8 x Belokokonna-1) x $ *(SANIISH-9 x* Belokokonna-2) and Tetrahybrid-4 $ (SANIISH-9 x Belokonna-2) x $ (SANIISH-8 x Belokonna-1). These complex hybrids were zoned and bred in production for more than 40 years.

The bivoltine breed TashSKHI-112 and monovoltine breed SANIISH-9 were used more than 35-40 years ago and are currently used for re-feeding [52;p.3-19].

The works of a number of scientists [108;p.3-7], [52;p.3-19], [123;p.8] provide methods of breeding productive breeds, and it is noted that when creating bivoltine breeds, it is necessary to pay special attention to the self-vitalisation of the grena.

§ 1.2. Genetic parameters as a theoretical basis for selection of mulberry silkworms

When creating breeds and hybrids for different feeding seasons and different climatic zones, breeders face great challenges: the created breeds and hybrids should have good grena revival, silk yield should meet the requirements of industrial enterprises, breeds should be resistant to diseases. Many scientists have studied these issues [151;p.624-631], [153;p.9-23], [156;p.53-59], [157;p.147-156].

To implement the set tasks and to avoid negative consequences arising in climatically unfavourable conditions of spring, summer and autumn, it is necessary to pay special attention to the study and testing of breeds and hybrids of mulberry silkworm in ecologically unfavourable zones with subsequent selection in extreme conditions of summer and late autumn.

Improvement of fodder quality is of no small importance. For this purpose, it is necessary to widely introduce into production highly productive zoned varieties of mulberry and promising candidate forms of varieties, for example, Jar-Aryk 1, 2, 3, 4, 5, 6 and others. Many of these varieties have already been released - Jar-Aryk 7,8, Holodnostepskaya-6, Zimostoyky, Mankentsky, Oktyabrsky, Pionersky, SANIISH-33, SANIISH-34, Surkh-tut, Holodnostepskaya-seedless, Uzbek, Jar-Aryk-2, Jar-Aryk-4, Jar-Aryk-5, Jar-Aryk-9, Jar-Aryk-10 [69;p.5].

With the help of introduction of promising varieties of silkworm into production, the fodder base of silkworm breeding is quickly strengthened.

A number of scientific researches are devoted to the problem of creation and introduction of new highly productive breeds and hybrids of silkworms. For example, [121;p.92]; [44;p.121-125]; [11;p.34]; [91;p.13]; [114;p.p.3-20], as a result of which after painstaking research and tests in production in recent years were recommended for introduction of breeds: Orzu, Yulduz, Ipakchi-1, Ipakchi-2, Uzbekistan-5, Uzbekistan-6, Turon-1, Ipakchi-1 x Ipakchi-2, Ipakchi-2 x Ipakchi-1. The hybrids AGU-112 x UzNIISH-9, UzNIISH-9 x AGU-112 were created for repeated foraging [52;p.3-19].

The role of breeding in intensification of agriculture is huge. At present, it is impossible to limit ourselves to the successes achieved in silkworm breeding in Uzbekistan. On the basis of biotechnology and genetic engineering it is necessary to strengthen the work on creation and introduction into production of new highly productive breeds, varieties and hybrids resistant to unfavourable environmental conditions and meeting the requirements of intensive technologies.

The idea that it is possible to develop better and more efficient methods of selection has long been expressed by scientists, [129;p.51-54]; [124;p.31]; [99;p.13-14]; [104;p.143], [107;p.5-6], [106;p.52-72]; [88;p.7], [89;p.1-49]. [46;p.22- 24], [58;p.3-100], [73;p.50], [76:p.4-66], [72;p.45], [74;p.45], [140;p.60], [141;p.96- 97], [142;p.87-118]. High-yielding breeds and hybrids of mulberry silkworm suitable for extreme conditions of maintenance were created.

For many years in the Republic of Uzbekistan it was not practised to create mulberry silkworm hybrids specifically for each region, because the plan of cocoon harvesting was fulfilled by one spring fattening [85;p.3], [86;p.3], [59;p.1-89].

Now there is a need for repeated worming. For the summer worming season, the breeds TashSKhI-112 and SANIISH-9, which are in the living collection of the Research Institute of Silk Breeding, were created [52;p.3.3- 19].

The mulberry silkworm (Bombyx mori L) has been reared at home for more than five thousand years. Through the use of various selection methods, the silkiness of the cocoon has increased by 8-10% of the original silkiness level. For many years of work on breed and hybrid breeding, breeders together with geneticists were able to increase the silk content in the cocoon from 13-14% to 23-24%, and the breeds offered to production by silk content reached $ 25%, $ 26%   [125;p.16-30];   [101;p.13]; [82;p.50]; [100;p.17]; [90;p.34];
[74;c.45].

The generally recognised classification of selection is reflected in the works of C. Darwin. All types of selection are aimed at identifying from numerous populations those individuals that contribute to the implementation of the intended goal of the research programme.

A number of scholars: [118;p.37], [119;p.64]; [144;p.210-245]; [145;p.84-112], [146;p.87-118];   [147;p.3-60], [143;p.21-40]; [92;p.1-28]; [148;p.95-118])

It was noted that with the cessation of selection at the final stage of reproduction, the productivity of mulberry silkworms decreases significantly. In the Research Institute of silkworm breeding at creation of lines and breeds for different zones and seasons of worm feeding individual and family selection are used: [47;p.12- 14]; [48;p.26-28], [49;p.25-27], [50;p.3-27], [51;p.23-24]; [134;author's note]; [54;p.3- 30], selection on the weight of caterpillar and pupa [63;p.59]; method of selection of male butterflies by maximum fertilisation of females: [93;p.20]; method of selection of butterflies by body size: [87;p.32]; selection of families by average egg weight: [115;p.5-8]; [128;p.1-8]; [80;p.9]; [18;p.12]. Several methods of selection of breeding families for disease resistance have been developed: [81;p.20]; [16:p.26-28]; [17;p.9-10]; [71;p.86-99]; [34;p.223-228]; [56;p.23-28]; [39;p.9], [40;p.22].

Turkmen scientists [82;p.50],   [83;p.33-34] have proved the connection viability of eggs with the viability of caterpillars, which confirms the data of NIISH scientists [124;p.31], [110;p.3-324], who in their studies noticed that the offspring of butterflies that emerged from cocoons in the last days are less viable, [98;p.10-11], [99;p.13-14] states that selection on the life span of male butterflies leads to increased viability of future offspring. [102;p.79] two heat tolerant breeds SANIISH 23 and SANIISH 24 were created under high temperature regime. [12;p.18-19], [13;p.66-68] offers a method of selection of breeding batches of cocoons by silkiness. [32;p.9-10], [33;p.21-25] studied the problem of feed payment.

Many works also reveal the problem of sources of variability for primary material. Many works [2;p.23], [3;p.18], [7;p.32], [4;p.9-61], [6;p.9-61], [4;p.9-61], [3;p.18], [3;p.18], [7;p.32], [4;p.9-61], [6;p.17], [8;p.125-131], [9;p.13-21], [10;p.102-105]; [42;p.184-185]; [84;p.28-30]. The works [31;p.10-15] are devoted to the establishment of the coefficient of inheritance of economically valuable traits of cattle in hot climate conditions.

In silk breeding, the first studies on establishing the heritability indices of breeding traits of mulberry silkworm were carried out [110;p.3-324], [107;p.5-6]. The efficiency of mechanised selection of breeding cocoons was predicted by the coefficient of inheritability of cocoons silkiness (h^2), equal to 0.5.

In the last period, studies on variability and inheritability of the main economically valuable traits were carried out: viability, cocoon weight, silk, length and thickness of cocoon thread, reproductive properties [1;p.81-91], sericin content in cocoon sheath [95;p.18], [94;p.9-20].

It is evident from the above that mulberry silkworm breeding is carried out in different directions using different methods and ways of selection. The possibility of selection is due to the variability in the populations. On the basis of these changes, the breeder-geneticist selects variation of individuals by body structure, physiological and other properties called variability. The source of variability of traits are differences in genotypes and environmental conditions. This has been clarified in studies [163;p.42-59], [164;p.992-997], [149;p.57-61], [150;p.478-486].

Publications [124;p.31]; [125;p.16-30]; [89;p.1-49]; [60;p.83-89]; [38;p.102-111]; [98;p.10-11], [125;p.16-30]; [89;p.1-49]; [60;p.83-89]; [38;p.102-111]; [98;p.10-11], [99;p.13-14]; [97;p.12- 52]; [64;p.27=29]; [109;p.9]; [65;p.151]; [79;p.28].

A more promising source of replenishment of breeding material is obtaining different kinds of mutations in insects, birds and animals, as indicated by [61;p.1-25]; [127;p.270]; [145;p.84-112], [146;p.87-118]; [37;p.63].

The great successes in fur farming (multi-coloured furs) are based on genetic spontaneous mutations in evolutionary processes and genetic methods for breeding these very rare mutations, as well as on genetic methods for analysing their effects, selecting mutants and creating forms resulting from the combination of two, three or even four mutations. It is no exaggeration to say that selection in fur farming now rests entirely on genetic principles. This is stated by [28;p.35-42]; [57;p.43-50]. Probably, the same can be said in the near future about the selection of various coloured forms of Karakul sheep.

Large scientists-geneticists B.L.Astaurov, N.K.Belyaev, N.P.Dubinin, M.I.Slonim, V.P.Efroimson made a great contribution to practical and scientific silk breeding. They published a number of genetic, important in theoretical and practical respect, works. Wide application in scientific and practical silk breeding of the developed new types of reproduction, such as parthenogenesis, gynogenesis, and androgenesis will allow to develop more effective methods of mulberry silkworm breeding.

Among a large number of received mutations it is quite probable to single out the mutation, the most suitable for selection and even easier to single out sex-linked visible recessive traits and volatiles [135;p.9], i.e. mutations arisen in X and sex chromosomes. They are detected in the offspring of heterozygous sex, as they appear in it in one dose or in the homozygous state.

More effective obtaining of modified individuals by ameiotic and meiotic parthenogenesis and androgenesis in mulberry silkworm is reflected in the works of V.A.Strunnikov [105;p.145]; [109;p.9], [107;p.5- 6].

§ 1.3. Influence of external environment on the manifestation of mulberry silkworm productivity traits

External conditions play a decisive role in the realisation of the genotype, in its fullest phenotypic manifestation. From genes and their primary protein products to the phenotypically mature organism, a number of processes take place that can be influenced to varying degrees by environmental factors and other changes. These complex relationships between genotype and environment are known as genotype response norms.

Unlike other insect species, the mulberry silkworm, as a strict monophage, feeds exclusively on mulberry leaves. The content of nutrients in leaves may vary depending on the variety of mulberry and agro-technique of plantation care. Thus, [116;p.113-116] experimentally revealed varietal differences in the main elements (in grams) of the food eaten by 100 caterpillars:

Silkworm variety	Total nitrogen	Protein nitrogen	Sugar	Raw fat
SANIISH variety 14	13,30	11,87	26,28	12,24
Variety Tajik seedless	15,72	14,07	35,55	15,36
In %% of SANIISH 14, taken as 100%	118,2	118,5	135,3	125,5

In special silkworm literature the influence of environmental conditions on growth, development and productivity of mulberry silkworm is covered in detail. [102;p.79],

[47;p.12-14] and other authors have shown the combined effect of feed quantity and quality, hygrothermal regimes and zonal peculiarities on the level of productivity of mulberry silkworm.

It is known that the mulberry silkworm is a strict monophage, it gets the necessary substances for growth, development and synthesis of silk from mulberry leaves. In years with unfavourable weather and climatic conditions, silkworm leaves do not accumulate the necessary amount of nutrients, for this reason silkworm caterpillars curl incomplete cocoons with reduced weight and silkiness.

A series of papers have been published on the use of different ingredients for enrichment of mulberry forage leaf.

Studies on the importance of ingredients in increasing productivity have also been carried out on the oak silkworm. Such works include experiments [15;p.13-17], which established the possibility of increasing the silk content in the cocoon.

[90;p.34], [91;p.13] states that creation of conditions for more complete realisation of genetic potential of mulberry silkworm with application of method of enrichment or enrichment of silkworm forage leaf with biologically active compounds opens possibilities to weaken negative influence of deficiency of nutrient elements and moisture in silkworm leaves, formed in unfavourable conditions with sharp weather-climatic deviations, frosts, precipitation deficit, excessively high background temperature during spring selection and breeding periods.

New varieties of mulberry have been developed by [43;p.27-29], [45;p.5], [66;p.33-35], [67;p.18-21], [68;p.26].

In our study, we propose to improve the productive qualities of mulberry silkworm by selecting appropriate mulberry varieties with large fleshy leaf plate, with optimal biochemical composition of forage leaf.

[16;p.26-28] on the basis of his research proposes to carry out selection by samples of breeding batches coming from breeding sites of drainage plants.

[18;p.12] suggests selection of parental pairs by silkiness and graininess of mulberry silkworm cocoons. Using this method, he obtained two lines - Uzbekistan 1 and Uzbekistan 2.

Certain genes manifest themselves most fully only in certain habitat conditions, which can be highly specific. Another group of genes manifests its action only in combination with other genes [35;p.30-44]; [120;p.13-104]; [138;p.1-22]; [139;p.62]; [137;p.356-375]. These are the dynamics of gene expression in the process of selection, which continuously introduces changes in the normal response of the organism.

At present, the demand for natural silk is increasing. Until the nineties, breeding of hybrid combinations of mulberry silkworm in the conditions of production on one spring feeding, quite satisfied the need for silk. At present, the task of repeated fattening is set. In conditions of hot dry summer not all breeds are able to fully show their genetic potential. Therefore, the need to breed and introduce mulberry silkworm hybrids for extreme environmental conditions, is increasing. Such hybrids can be used both in areas with difficult natural conditions and for repeated foraging in the southern regions of Uzbekistan.

The economic benefits of spring and re-feeding can be seen in the example of the Jar-Aryk experimental farm (silk breeding institute), where after the spring foddering there is a lot of uncut fodder leaf left after the spring foddering. In recent years in Tashkent and Tashkent oblast a lot of unused leaf mass has been left. In the majority of farms also annually remains uncut leaf, as a result of which in the following years silkworm becomes small-leaved, seed. Therefore, planning of intermediate and summer fattening in the conditions of market economy will bring double benefit: additional cocoons harvest and strengthening of fodder fund in the field.

The success of summer and autumn feeding depends largely on the provision of quality grenache. Since grena is prepared in spring for summer, it was necessary to develop effective ways to remove diapause. In this direction, the first studies were carried out by N. Koltsov. He studied the effect of high temperature, various organic and inorganic acids, alkalis, drugs, oxidising agents, iodine and some fixing liquids on unfertilised grena of monovoltine species. High temperature and iodine proved to be the most successful activators, but despite this only 15 caterpillars were obtained. Studies on the effects of high and low temperatures on mulberry silkworm development continued. They add to the already large list of works in which temperature effects have been used with great success to control the sex of the mulberry silkworm. Thus, for example, meiotic and ameiotic parthenogenesis- [19;p.136- 140], androgenesis- [27;p.277-280], [26;p.3-21]; polyploidy- [24;p.130-141], [23;p.344-367]; control of volatility and elimination of diapause- [20;p.3- 21], [22;p.77-81], [19;p.136-140], [25;p.9]; [130;p.395-424], [131;p.28]; lifetime decontamination from pebrina spores- [96;p.54-63]. In his works B.L.Astaurov rightly noted that the intervention in the process of mulberry silkworm life activity should be carried out by the factors encountered by the organism in nature. This position is well confirmed by the results of our experiments. Viable and less viable grena do not differ significantly in resistance to factors not met by the organism in nature, against the action of which in the process of evolution have not developed special adapted reactions, significantly reducing their harmful effect.

In white-coloured high-yielding breeds and hybrids of silkworms, cocoons yield decreases when the beginning of fattening is late, mainly due to reduction of cocoon weight. Thus, for example, in the experiment conducted by G.V.Priyezzhev in Tashkent district, the average weight of cocoons of the hybrid SANIISH-8 x SANIISH-9 was 2.40 g at the beginning of feeding, after the appearance of the third leaflet on the silkworm, and 1.84 g when delayed by 14 days.

The effect of external environment (temperature influence) was clearly observed in the conditions of repeated fattening, when feeding the same seedling of the hybrid SANIISH E-1 x Belokokonnaya 1 and its reverse combination, in 1960 received from one box of seedling of the above-mentioned hybrid from 87 to 74 kg of cocoons. In the same year in the farm located nearby, feeding the same hybrid on intermediate

feeding, received an average yield of 40.3 kg of cocoons from 1 box of grena. Also positive results can be seen in the works: [55;p.11]; [122;p.8]; [75;p.35]; [132;p.3- 42]. The works of many researchers on the impact of various factors, in particular, temperature regime on biological and economic characteristics of insects and animals are known. Therefore, the issue of high-temperature effect on animals and resistance of organisms to extreme conditions remains relevant, and carrying out works in this direction deserves approval. Scientists [84;p.28-30]; [136;p.16-17]; [126;p.16]; [53;p.11-15]; [117;p.48-49]; [62;p.10] and others,
have devoted their research activities to studying the effects of temperature exposure on the vital processes of farm animals.

Most insects do not react in the same way to external changes in temperature and humidity and to other factors. The silkworm in its development reacts quickly to such changes. According to the studies conducted by E.F.Poyarkov [96;p.54-63], it can be seen that under the action of high temperatures on mother pupae during their transformation into a butterfly, purification from many infectious diseases takes place, which contributes to the recovery of the next generation of silkworms. The works [29;p.42], [30;p.11-12] proved that the change of temperature regime in the period of spring incubation affects the vivacity of the gren and other economic indicators. From the works [14;p.32], [15;p.13-17] it can be seen that by maintaining the optimal hygrothermal regime, the expected productivity of silkworms is obtained, and if this regime is violated, the weight of the cocoon and its shell decreases, as well as the viability decreases.

The Republics of Central Asia are characterised by high temperature and low humidity, so rearing of mulberry silkworm caterpillars in these regions requires selection for spring and intermediate, summer and summer-autumn rearing of specially prepared premises with optimal rearing regime adopted for white horse breeds and hybrids and highly nutritious fodder.

The creation and selection of special breeds and hybrids of mulberry silkworms plays an important role for the success of intermediate, summer and summer-autumn fattening. Many years of research have been carried out in this regard [102;p.79]. The author created heat-resistant breeds SANIISH-23 and SANIISH-24 and their hybrids were zoned in 1971 in Turkmenistan. The main parameters of the direct hybrid SANIISH-23 x SANIISH-24 under the conditions of laboratory tests: cocoon weight - 2.31 g, cocoon yield from 1 box - 99.6 kg, total length of cocoon thread - 1283 m, metric number - 3150 units. These indicators met the needs of the silk industry of Turkmenistan.

The majority of producers in those times argued that in case of repeated fattening the volume of spring and intermediate fattening should be much larger than summer fattening, because intermediate fattening is carried out with less damage to the fodder base of the subsequent year and provides higher cocoons yield. The use on repeated fattening of grenae prepared in the previous year creates conditions for high-quality preparation of grenae at grenade plants due to reduction of a number of production processes and increase of grenae yield from breeding cocoons.

The effect of external conditions was reliably manifested in the production conditions. According to the researchers [55;p.11] it can be seen that the intermediate and summer fattening yields are low, which is not economically favourable for silk production. But, nevertheless, the yield from intermediate fattening was 10 kg more than that from summer fattening. These fattening operations were carried out more than 60 years ago. During this time 15 hybrids of mulberry silkworm intended for spring season have been zoned in silk breeding. For summer season A.A.Sheveleva and N.V.Shurshikova created hybrids TashSKHI-112 x SANIISH-9 and SANIISH-9 x TashSKHI-112. Taking into account the need of industry in natural silk R.K.Kurbanov and M.R.Kurbanova (NIISH) created Line-12 and Line-40 for extreme summer conditions, components of hybrids Yozgi-1 and Yozgi-2, which were zoned in 1997.

Silkworm breeders know that it is impossible to obtain quality raw silk without quality silkworm leaf forage. Silkworm breeders have done a lot in the direction of breeding high-yielding varieties and hybrids of mulberry.

It is known that the mulberry silkworm, being a strict monophage, gets the necessary substances for growth, development and silk synthesis from mulberry leaves. In years with unfavourable weather and climatic conditions, mulberry leaves do not accumulate the necessary amount of nutrients, which is why silkworm caterpillars curl incomplete cocoons with reduced weight and silkiness.

Silkworm breeders have searched for the most nutritious and long non-fading fleshy silkworm varieties for caterpillar foraging on off-season summer and autumn forages. Nine yielding varieties were recommended for use in off-season foraging. [32;p.9-10], [33;p.21-25] proved that the increase of mulberry silkworm cocoons yield can be achieved with the timely start of foraging and proper selection of mulberry varieties.

In our study, we propose to seek to improve the productive qualities of mulberry silkworm by selecting appropriate mulberry varieties with large fleshy leaf plate, with optimal biochemical composition of forage leaf. A special search of scientific literature was conducted to study, evaluate and recommend for successful fattening of mulberry varieties with better characteristics.

The Central Asian Republics differ in their climatic peculiarities from the Eastern and South-Eastern countries by high temperature and low air humidity. Therefore, rearing of mulberry silkworm caterpillars requires selection of certain varieties of silkworms, careful preparation of special premises providing optimal rearing regime.

It is well known that the Central Asian countries of Turkmenistan, Kyrgyzstan, Tajikistan, Kazakhstan, whose climate in the spring worm-feeding season is almost similar, have so far fulfilled the plan of cocoon harvesting by carrying out one feeding, feeding mulberry silkworm hybrids curling cocoons of white colour. When re-feeding, the above-mentioned countries differ sharply from each other in climatic conditions. Our country is considered to be the most suitable for re-feeding. We have more than ten high-yielding and highly productive varieties and hybrids of mulberry introduced in production for this purpose.

But, unfortunately, the work on wide introduction of released varieties of mulberry is very slow.

Newly zoned hybrids of mulberry silkworm are also poorly spread in production, in particular, hybrids with participation of sex-regulated breeds: C-13 x C-14, C-14 x C-13, Mech-1 x Mech-2, Mech-2 x Mech-1, clonal 100% pure hybrid of one direction - 51.40 pk x Sovetskaya 5, as well as unlabelled breeds Ipakchi-1, Ipakchi-2 and their hybrids Ipakchi-1 x Ipakchi-2, Ipakchi-2 x Ipakchi-1. These hybrids were created in the late nineties, passed three-year state and two-year production tests, after which their zoning in the country was allowed. Among the above-mentioned hybrids, only Ipakchi-1 x Ipakchi-2, Ipakchi-2 x Ipakchi-1 are widely introduced.

As can be seen from the brief literature review, much research work has been devoted to the development of new breeds and hybrids of mulberry silkworm, the development of new varieties of silkworm and its hybrids, and the development of new methods of mulberry silkworm breeding. However, to date, the amount of cocoons produced is not enough to provide the textile industry with enough raw materials to ensure that the raw materials are of high quality. Therefore, in recent years, special attention has been paid to the quality of raw silk.

Complex assessment of suitability of varieties for foraging of mulberry silkworm caterpillars in ecologically difficult conditions of repeated foraging was carried out in NIISH. The evaluation, taking into account the indicators of unwinding, raw silk yield, production length of cocoon thread and its metric number (fineness) showed that the best varieties were the best in terms of manifestation of technological traits:

Jar aryk 4,

Tajik seedless,

Oktyabrsky.

From the analysis of cocoons on technological indicators obtained under unfavourable conditions of late-spring foddering, the conclusion is that varietal features of forage influence not only on the manifestation of biological traits, but also on the qualitative characteristics of unwound silkworm.

Summarising the results of the search of scientific literature reflecting experiments on the use of mulberry of different varieties to increase the yield of mulberry silkworm cocoons, we consider it possible to recommend feeding caterpillars in extreme conditions of Karakalpakistan with mulberry leaves of the released varieties Tajik seedless, Oktyabrsky and promising variety Jar Aryk.

4, and to use these varieties in feeding thin-silk breeds and hybrids.

CHAPTER II

RESEARCH MATERIAL AND METHODOLOGY
§ 2.1. Choosing a research direction

The research was conducted in the laboratory of genetics and breeding of mulberry silkworm at NIISH and at Nukus Agrarian University from 2011 to 2017.

The choice of research direction is dictated by the need to provide industrial silkworm breeding in Karakalpakistan with new, highly productive, resistant to unfavourable environmental conditions of this region, breeds and hybrids of mulberry silkworm. Such orientation of the work determines the corresponding methods of research: analysis, selection, approbation and selection.

The breeds and lines contained in the collection of mulberry silkworms of NIISH [76;p.4-66]), as well as hybrids created in the laboratory of genetics and breeding of mulberry silkworms of NIISH [109;p.9], [113;p.28-31] were used in this work. The components of these hybrids are sex-labelled at the egg stage of breeds C-5, C-5ngl, C-7, C-9, C-10, C-12, C-13, C-14, sex-labelled at the caterpillar stage of breeds Mechnaya 1, Mechnaya 2, as well as sex-labelled breeds SANIISH 30, Asaka, Markhamat, Atlas, Margilan, Ipakchi 1, Ipakchi 2 and parthenogenetic clones.

§ 2.2. Methods of breeding work with breeds, lines, hybrids of mulberry silkworms

Work with breeds was carried out according to "Basic methodological provisions of breeding work with mulberry silkworm", [4;p.3-16], in which small changes were made taking into account genetic features of sex-labelled breeds and parthenogenetic clones. Thus, each family of sex-labelled breed at the grena stage was incubated separately by sex. 110 males and 110 females (dark grena, light grena) were counted from each family and fed together. At the grena stage, families were rejected for reproductive traits, sex ratio and percentage of grena revival. Families with grena revival below average were not selected for fattening. At the second instar, families and repeats of 220 caterpillars each were formed, and hybrids intended for laboratory testing were counted at 150 caterpillars in a three-roller repeat. At the caterpillar stage, culling was done in case of heterogeneous development and low viability of caterpillars. Families were analysed using a sample of 30 cocoons (15 females and 15 males). Families with low silkiness, cocoon weight and shell weight were rejected. Individual analysis of cocoons by breed was carried out separately for females and males. According to the results of individual analysis, cocoons with large shells, high silkiness and breed-specific shape and graininess were selected for preparation of clutches of source material. The best individuals were crossed with the best individuals by the outbreeding method. The remaining cocoons of breeding families after individual selection were used to prepare super-elite and hybrid grens. Production hybrids were reared with one to three boxes of grena under production conditions.

Technological analyses of breeds and clones were carried out using a sample of 300-400 cocoons made up of cocoons from all replicates or reared families. The failure of butterflies to emerge from cocoons was determined using a total sample of 250-300 cocoons.

In the process of selection and breeding work with families, breeds and clones, the following indicators were determined:

- number of eggs in the clutch (pcs);
- presence of physiological marriage (%);
- average weight of one clutch and one egg (mg);
- grena revival in clutches or samples (%);
- caterpillar viability (%);
- presence of grouse (%);
- duration of the caterpillar stage (days);
- silk content in live cocoons (%);
- average mass of a live cocoon (g);
- average mass of silk sheath (mg);
- failure of butterflies to emerge from cocoons (%);
- raw silk yield (%);
- silk products (%);
- cocoon unwindability (%);
- length of continuous unwinding (m);
- total production length (m);
- metric number of the thread.

§ 2.3 Reproduction technique for parthenogenetic clones

Parthenogenetic clones for the study were reared in replicates (2-16 replicates each). In each replicate there were 220 pieces.

Activation of clones was carried out according to the method of B.L.Astaurov by treatment of unfertilised grena from 1-5-10 moths with hot water at a temperature equal to 46^0 C for 18 min [27;p.277-280], [26], [25;p.9] with subsequent storage for 3 days at a temperature of $16-17^0$ C and humidity 90-95%. Then the grena is stored at the usual temperature accepted for storage of white horse breeds of mulberry silkworm.

Reproduction of parthenogenetic clones was carried out according to the method of B.L.Astaurov [18;p.240]:

1. The contents, consisting mainly of ovariole, are squeezed out of the torn off abdomen of the butterfly and gently rubbed through a tulle sieve with fingers under a stream of water. The eggs, having passed through the sieve, fall to the bottom of the glass vessel provided. Another part of the abdomen remains on the sieve surface.

2. The washed grena is spread in a thin layer on filter paper and stored at $16-17^0$ C for 10-12 hours.

3. The grenache is then placed in a double-layer gauze knot, loosely tied so that the hot water freely washes over the whole grenache. A parchment label, signed with a simple hard pencil or red biros, is placed in each knot. The heating is done in water at 46^0 C and lasts 18 minutes. The warmed grenache is immediately transferred to room temperature water (25^0 C) for 5-6 minutes.

4.   The heated and cooled grenin is spread out in a thin layer on filter paper and dried under the jet of a fan, which should be so removed that the air jet does not carry the grenin off the filter paper.

5.   The dried grena is transferred to a room with a temperature of 16-17^0 C and humidity of 90-95%, where it is stored for 3 days.

6.   After this time, the grena is transferred to a 250C room and undergoes normal estivation or treated with HC1 if diapause is to be interrupted.

After microanalysis, the grena was poured into parchment unemulsified bags and placed in a cold room for storage until the next foraging.

Eggs activated in this way develop normally, and only females strictly repeating the genotype and all features of the mother-parent emerge from them. Simple heating of unfertilised eggs in water 46^0 C for 18 min, allows to get as many parthenogenetic butterflies as desired, always the same females.

Parthenogenetic clones AIC and 9PK were crossed for six years with the thin-silked breed Ya-120, with which selection for improvement of productive properties and selection for cocoon sheath granularity in order to thin the cocoon thread were carried out during all these years. Hybrids APK x Ya-120, 9PK x Ya-120 were reared in 3^x repetitions on 200 caterpillars in each according to the generally accepted technology of keeping white-cocoon breeds of mulberry silkworm. During the work all necessary reproductive, biological and technological parameters of hybrids were recorded and analysed.

Figure 1.3.2 shows the scheme of obtaining ameiotic parthenogenetic females of mulberry silkworm. Figure 1.3.2 shows female butterflies of parthenogenetic clone.

The mulberry silkworm has 56 chromosomes in its genotype. One pair of chromosome is sex chromosome, the other 27 pairs are autosomal chromosomes (Fig.2.3.1).

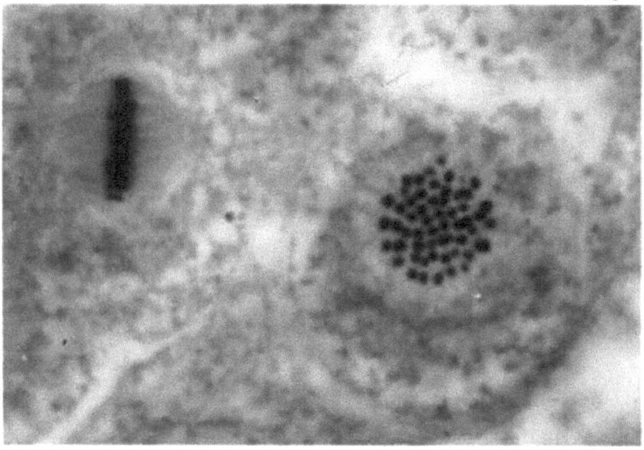

Fig. 2.3.1 Chromosomes of the mulberry silkworm at the Anaphase stage

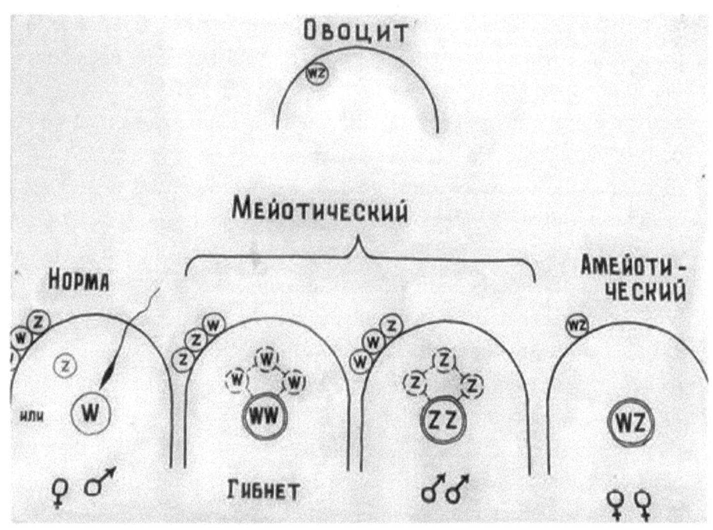

Fig.2.3.2 Scheme of obtaining amaeiotic parthenogenetic females of mulberry silkworms

Fig. 2.3.3 Constant uniform female butterflies of a parthenogenetic clone

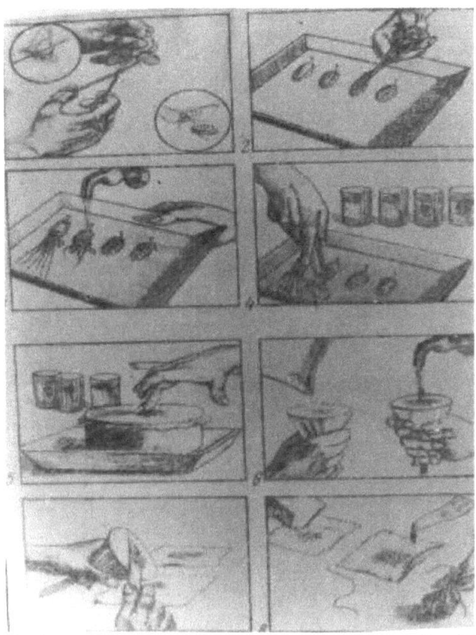

Fig. 2.3.4: Technique for extracting and preparing unfertilised mulberry silkworm eggs for subsequent thermoactivation and thermal artificial parthenogenesis.

Figure 1.3.4 shows how an unfertilised gren is prepared for reproduction the following season. After washing, the grena is tied in knotted gauze and heated at 46^0 18^1 . After resting for 12 hours at 16-17^0 C with 85-89% humidity, it is stored at normal temperature like bisexual breeds.

§ 2.4 Methodology for obtaining translocations in mulberry silkworms

To increase diversity in mulberry silkworm populations and to produce new mutations and genetically modified organisms, the silkworm genome is subjected to various influences.

For example, by gamma ray treatment of the mulberry silkworm genotype, a translocation of the dominant gene +W2 from autosome 10 to the W chromosome was obtained [106;p.52-72], [107;p.5-6], [103;p.50] (Figure 2.4.1).

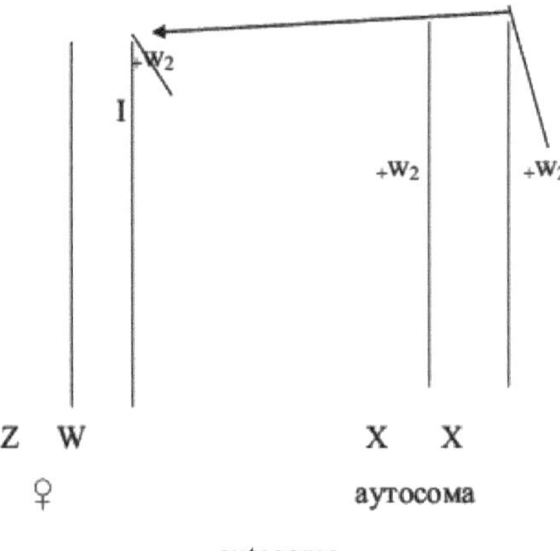

autosome

Fig. 2.4.1 Scheme for obtaining a genetic translocation

As a result of this mutation, the silkworm butterfly lays dark and white grens. From the dark eggs come to life ♀♀, , from the light eggs come to life males ♂♂. .

Thus, for example, breeds C-12, C-5, C-9, C-13, sex-labelled with gene W_2 , producing straw-coloured grena (males); breeds C-10, C-14, labelled with gene w_3, which causes dark brown colour of grena (males); breeds B-2 W_5 , C-6 W_5 , labelled with gene W_5 , with dark brown colour of eggs (males) and others were obtained. For our research the breeds C-5, C-10, C-12 are of the greatest interest. The peculiarity of the method of selection and breeding work with sex-determined breeds at the grena stage is that after microanalysis of butterflies, each clutch of sex-labelled breed is divided by colour, the number of dark and light eggs is counted, then only clutches with the ratio of eggs 50% ♀♀, : 50% ♂♂. are selected for further work, the rest are rejected. During rearing each family is formed of 110 females and 110 males, or females and males are reared separately to avoid further division of pupae by sex.

Families of the breeds Sword 1, Sword 2, labelled by sex at the caterpillar stage, are divided by sex at the fifth instar into males (white-milk caterpillars) and females (caterpillars with masks and half moons) with counting of caterpillars of each sex.

Only families with a sex ratio of 50% ♀♀, : 50% ♂♂. are docked for pedigree crosses. After separation by sex,

caterpillars are placed on different shelves and curl cocoons separately.

Figure 2.4.2 shows a photograph of a sex-labelled breed grena.

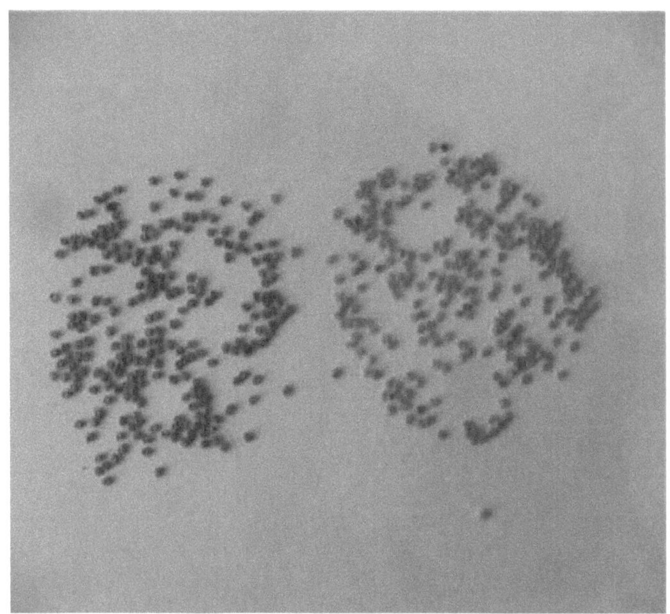

Fig.2.4.2. The grena of a sex-labelled breed divided by colour into females (dark grena) and males (light grena)

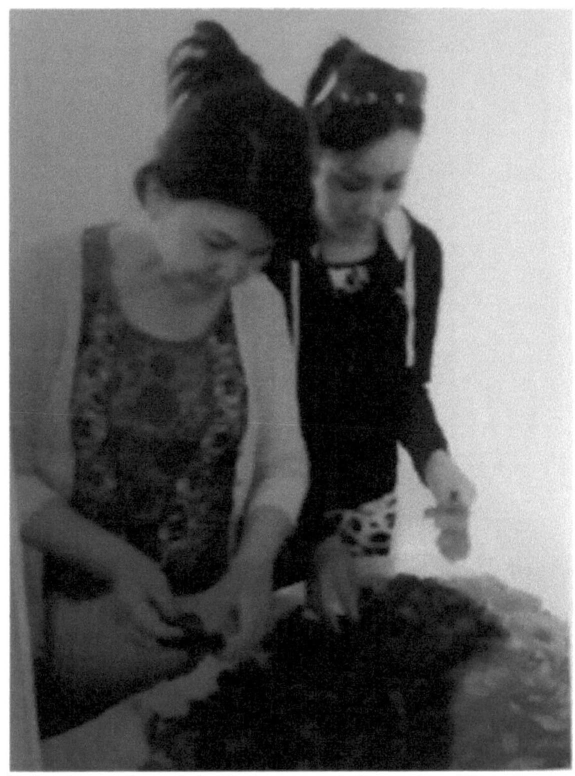

Fig.2.4.3.Revitalisation of sex-labelled grens at the caterpillar stage of the breed

Fig.2.4.4 Division of caterpillars of the breed labelled by sex at the caterpillar stage into white (males) and patterned (females) caterpillars

Fig.2.4.5 Removal of cocoons from cocoons

§ 2.5 Selection methodology on behavioural activity

Known methods of increasing viability, silkworms, increasing the weight of silk sheath, percentage of grena revival, yield are based on labour-intensive selection of silkworms at the stage of grena, caterpillar, pupa, butterfly with the most valuable indicators. These methods due to lack of selection accuracy do not exhaust the possibilities of additional increase of silkworm productivity. Therefore, in order to save money, our employees are looking for other methods of preserving at the achieved level of valuable indicators of sex-labelled breeds. One of such methods is selection by motor activity [72;p.45-51]. According to this method, at the beginning of each age, and at the moment of exit from the cocoons of butterflies, the most mobile and active males at crossing are selected. For this purpose, before revival of the grena during the period of its whitening, a mulberry leaf is put into the bag with the grena in such a way that it did not touch the eggs and was at a distance of 1-1.5 cm from them. During revival, the most active and mobile, and therefore the most viable caterpillars are the first to climb onto the leaf. In 2 hours after the start of reanimation, the leaf with a sufficient number (about 300 caterpillars) is removed and further, in order to save feed, the caterpillars are fed until the third instar in parchment bags.

For this purpose, the most active caterpillars are placed in perforated parchment bags of 20 cm x 30 cm and fed with chopped young mulberry leaves 3 times a day (instead of 8-10 times according to the usual method). The bags with caterpillars are placed under moistened cover. Formation of families or repetition is made in the third instar, selecting the most active caterpillars during counting. Culling of passive caterpillars during reanimation and counting leads to purification of material from individuals carrying recessive lethal genes in genotypes at different stages. Then the caterpillars are reared in the usual way accepted for rearing white horse breeds.

To select mobile male butterflies, females are placed on one side of the brooding bed and males are placed on the other side of the bed at a distance of 30-25 cm. The most mobile males begin to move actively towards the females. After 10-15 minutes, all unstirred males are removed from the bed. Thus, the offspring are left by the most viable individuals.

Selection by motor activity is used in the last 10 years to maintain breeds C-13, C-14, divided by the colour of eggs into females (dark grena) and males (light grena) and divided by sex by the colour of caterpillars Meechennaya-1, Meechennaya-2. The use of more mobile caterpillars and the most active male butterflies contributes to the health improvement of breeding material as a whole and makes it possible to maintain the performance of breeds and hybrids at the achieved level.

Figure 2.5.1 shows the overall research design.

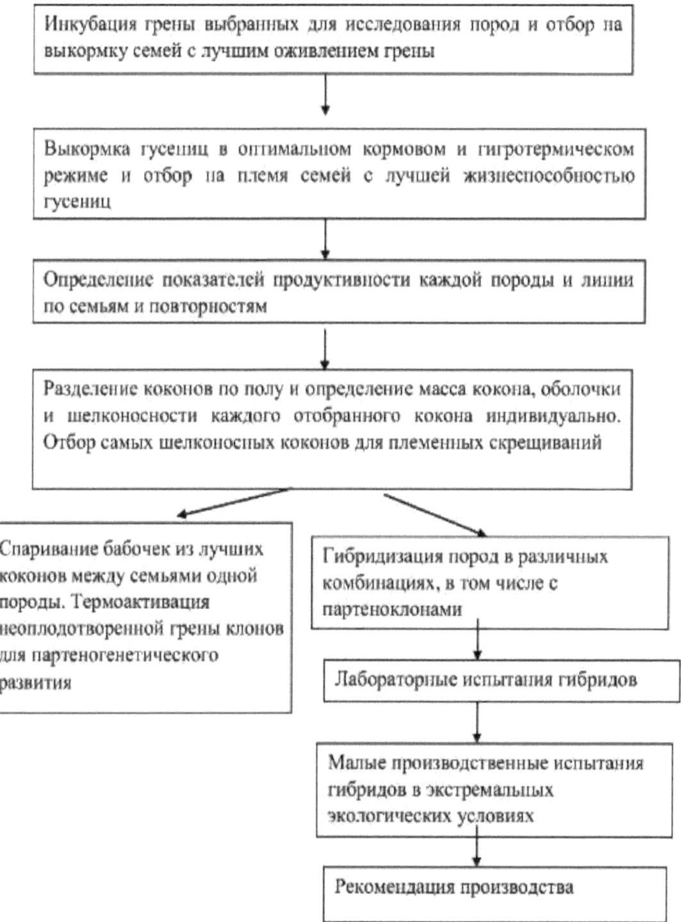

Incubation of grens of the breeds selected for the study and selection of families with the best gren revitalisation for fattening

Feeding of caterpillars in optimal feeding and hygrothermal regime and selection of families with better caterpillar viability for breeding
Determination of productivity indicators of each breed and line by families and repetitions
Separation of cocoons by sex and determination of cocoon weight, sheath and silkiness of each selected cocoon individually.
Selection of the silkiest cocoons for breeding crosses
Mating of butterflies from the best cocoons between families of the same breed. Thermoactivation of unfertilised clone genes for parthenogenetic development
Hybridisation of breeds in various combinations, including with parthenoclones
Laboratory testing of hybrids
Small-scale production trials of hybrids under extreme environmental conditions
Recommendation for production

Fig. 2.5.1 Schematic diagram of experiments

CHAPTER III

OBTAINED RESULTS

§ 3.1 Characterisation of source rocks

The thesis work involved conventional and sex-labelled breeds at the egg and caterpillar stage.

As it is known, silkworm grena can be stored up to eight months in the refrigerator at $t=+2+3^0$. With the continuation of storage time, deteriorates the vivacity of grena and viability of caterpillars in April-May by 3-6%, in June-July by 12-18%, in August not 55-60%. Therefore, in order for silkworms to retain their biological characteristics, breeders need to reproduce them annually using selection. Without selection, performance deteriorates rapidly. It will take 4-5 years to raise the results achieved by the labour of breeders.

For the work, clades of source material were prepared annually and characterised in Table 3.1.

Table 3.1.

Characterisation of initial clutches by species (2012)

Name of material	Number of pieces			Average number of eggs per clutch	Weight		% physical defects
	a clutch.	display on off.	culling, on culling.		of masonry, g.	of one egg, mg	
C-5	164	107	24	571	349	0,61	4,67
C-5ngl	84	46	21	545	292	0,53	6,03
C-7	75	60	22	620	346	0,55	1,58
C-9	82	71	25	600	335	0,56	2,91
C-10	132	96	41	561	308	0,54	4,59
C-12	39	37	33	518	275	0,53	10,9
C-13	310	144	62	538	292	0,54	4.94
C-14	194	107	54	543	278	0,51	2,51
Sworn 1	330	148	48	648	363	0,56	3,04
Sworn 2	323	172	56	656	368	0,56	1,56
SANIISH 30	131	97	22	665	356	0,53	2,77
Asaka	62	62	26	614	348	0,56	4,80
Marhamat	71	71	30	614	337	0,54	4,2
Atlas	95	95	39	608	365	0,60	4,1 0
Margilan	161	93	30	677	358	0,52	4,10

As can be seen from the table above, the highest number of eggs is observed in unlabelled breeds such as Asaka, Markhamat, Atlas, Margilan and SANIISH 30, as well as in caterpillar stage labelled breeds Mechnaya 1, 2. In the breeds labelled at the egg stage, the lowest number of eggs in the clutch and a greater number of physiological defects were observed. This is explained by the genetic peculiarity of these breeds having genetic rearrangements in their genomes. As for the clutch weight, it varies proportionally with the average number of eggs in the clutch. The mass of one egg varies slightly from 0.51-0.61 mg, depending on the breed.

It should be noted that at all stages of mulberry silkworm development the work on breed improvement is carried out. Thus, during the estivation period, clutches are culled according to the following indicators: clutch weight, weight of one egg, percentage of physiological defect and sex ratio in the labelled breeds. This is reflected in the column "number of clutches prepared" and "number of clutches selected for incubation". Significant differences in the numbers indicate that a rigorous selection process is in place at this stage (Table 3.1.).

All breeds are improved by culling poorly reanimated families. The best families are taken for fattening, or replicates are formed from the best families in terms of gregariousness. This is reflected in Table 3.2.

Table 3.2.

Livability of the grena of the breeds used in the study

(cf. for 3 years)

No. of items	Name of breed	Number of normal eggs (pcs)	Number of unhatched eggs (pcs)	% of grena revitalisation
1	2	3	4	5
1.	AGU-112	633	19	96,9
2.	UzNIIISH-9	654	30	95,4
3.	Sworn 1	650	20	95,9
1	2	3	4	5
4.	Sworn 2	670	21	96,4
5.	SANIISH-30	586	14	97,6
6.	Sovetskaya-5	584	25	95,7
7.	Sovetskaya 5 (proz.gus)	519	39	92,4
8.	Sovetskaya-10	496	35	92,9
9.	Sovetskaya 12	516	60	88,3
10.	Sovetskaya 13	525	30	94,2
11.	Sovetskaya 14	528	29	94,5

It is clearly seen that all breeds are characterised by good grena revitalisation: 88.3-97.6%.

Biological indicators are also quite high.

Table 3.3.

Biological indicators of the breeds used in the study

(3-year average)

No. of items	Name of breed	Caterpillar life, %	Mean weight		% silk sheath
			Cocona, g	Shell, mg	
1	AGU-112	80,0	1,62	339	20,9

2	UzNIIISH-9	90,0	1,70	389	22,9
3	Sword 1	85,2	1,68	384	22,9
4	Sword 2	87,0	1,61	367	22,8
5	SANIISH-30	90,9	1,68	380	22,6
6	Sovetskaya 5	87,0	1,71	3768	22,2
7	Sovetskaya 5 prozr. gus.	88,8	1,52	319	21,0
8	Sovetskaya 10	92,1	1,47	349	23,8
9	Sovetskaya 12	88,5	1,67	397	23,9
10	Sovetskaya 13	84,8	1,58	389	24,6
11	Sovetskaya 14	89,5	1,65	387	23,5

As can be seen from Table 3.3, all breeds have good caterpillar viability: 84.8-92.1%. Breeds labelled by sex at the caterpillar stage are medium-skinned but high-shelled: Sovetskaya 13-24.6%, Sovetskaya 1223.9%, Sovetskaya 10-23.8%. Breeds labelled by sex at the caterpillar stage Mechenaya 1, Mechenaya 2 also have good silkiness: 22.9%, 22.8%.

In a separate experiment in 2012, biological parameters and coefficients of variation of some of the breeds used in the work were investigated. The results are summarised in Table 3.4.

Table 3.4 Biological indicators and coefficients of variation of the studied breeds (2012)

Breeds	Caterpillar viability, %		Cocoon weight, g		Weight of shell, mg		Silkiness, %	
	$X \pm sJ$	c_v		c_v		c_v		c_v
C-5	91,8±1,2	15,1	1,59±0,02	4,1	333±4,4	6,9	20,8±0,2	5,2
C-14	92,5±3,3	14,8	1,48±0,21	6,6	346±6,4	4,8	23,6±0,2	4,5
blink	94,3±2,5	13,0	1,75±0,02	5,5	404±0,7	0,9	23,1±0,3	6,9
PS-5	97,7±1,5	13,2	1,45±0,02	4,2	276±6,6	6,7	19,3±0,7	10,3
SANIISH 8	98,5±1,7	2,5	1,46±0,02	4,1	276±7,0	6,6	19,1±0,3	4,5
Я-120	98,7±1,2	2,1	1,65±0,04	6,3	411±11,3	7,7	25,3±0,5	5,5

As can be seen from Table 3.4, viability of caterpillars of sex-labelled breeds C-5, C-14, MG, PS-5 was slightly lower than viability of unlabelled breeds SANIISH 8 - 98.5%, Ya-120 - 98.7%. Sex-labelled breeds differ from normal breeds by the presence of translocation in their genomes. Therefore, these breeds are more sensitive to any changes in housing conditions. As it is known from numerous researches of Strunnikov V.A. (1969, 1987, 1994) in good experimental conditions biological traits of sex-labelled breeds are at the same level as in normal material, but in bad ecological conditions material with genetic changes behaves somewhat worse. High coefficients of variation 15.1; 14.8; 13.0; 13.2 of sex-labelled breeds testify to the high variability of such indicator as caterpillar viability and indicate the possibility of further selection for this trait.

The studied breeds belong to breeds with average cocoon (1.45-1.75 g) and sheath (276-411 mg) weight (Table 3.5). Further selection will be directed towards increasing the silkiness of cocoons.

Table 3.5 summarises the biological parameters of the breeds under study.

Table 3.5.

Biological parameters of sampled families and cocoons of the studied breeds (2012)

Breeds		Caterpillar viability, %	Weight		Silko-nosiness, %
			coc.g	obol. mg	
C-5	Feeding families	91,8	1,59	333	20,8
	Tribes, families.	95,6	1,58	340	21,5
	Tribal.kok.	-	1,56	348	22,3
C-14	Feeding families	92,5	1,48	346	23,6
	Tribes, families.	94,0	1,47	350	23,8
	Tribal.kok.	-	1,49	363	24,4
MG	Feeding families	94,3	1,75	404	23,1
	Tribes, families.	95,8	1,65	396	24,0
	Tribal.kok.	-	1,68	409	24,3
PS-5	Feeding families	97,7	1,45	276	19,3
	Tribes, families.	98,0	1,46	285	19,6
	Tribal.kok.	-	1,46	291	19,9
SANIISH 8	Feeding families	98,5	1,46	276	19,1
	Tribes, families.	98,9	1,46	288	19,7
	Tribal.kok.	-	1,48	396	20,0
Я-120	Feeding families	95,7	1,65	411	25,3
	Tribes, families.	98,0	1,65	426	25,8
	Tribal.kok.	-	1,69	438	25,9

Table 3.5 shows that families with silkiness of 21.5% were retained for breeding, while cocoons with silkiness of 22.3% were allowed for papillonage. In breed Ya-120, silkiness of reared families was 25.3%, breeding families 25.8%, and grena was obtained from butterflies from cocoons with silkiness 25.9%. Similar selection was carried out in breeds C-14, MG, AS-5, SANIISH 8 (Table 2.2.8).

In total, about 3,500 cocoons were analysed individually. Of these, about 1,800 cocoons were allowed to produce breeding grens, i.e.

54%. This selection intensity is expected to result in improved performance of the breeds under study.

Table 3.6

Technological indicators of breeds (3-year average)

Name of breed	Total of one dry cocoon, g	Yield, %		Metric number of cocoon thread,units	DNRCN	Unwinding - bridging	Production yarn length, m
		Silk-raw material	silk				
AGU-112	0,652	39,93	45,43	3327	703	87,91	946
UzNIIISH-9	0,674	36,04	47,75	3594	880	89,22	1141
Sworn 1	0,642	42,74	49,11	3522	805	87,29	1017
Sworn 2	0,681	43,34	49,51	3385	839	87,53	1156
SANIISH 30	0,654	42,98	49,07	3136	733	87,56	966
Sovetskaya 10	0,618	42,67	50,01	3436	794	85,32	1016
Sovetskaya 13	0,633	42,41	48,60	3548	754	88,18	1011
Sovetskaya 14	0,655	42,00	48,75	3073	687	87,06	931
Ipakci 1	0,641	41,28	47,89	3270	803	86,20	979
Ipakci 2	0,688	42,65	48,71	3033	764	87,56	959

used in the study are not inferior in silk thread quality to conventional breeds. For example, the metric number of Mechennaya 1 reaches 3522 units, while that of Sovetskaya 13 reaches 3548 units. The production length of thread of Mechennaya 2 is 1156 metres, while that of Sovetskaya 10 is 1016 metres. The UzNIISH 9 breed, which is a component of a hybrid intended for re-sprouting, attracts attention: production yarn length 1144 m, metric number 3594 units, unwinding rate 89.22%, DNRKN 880 m. All breeds have rather high yields of raw silk and silk products.

In sex-labelled breeds using gamma rays, the dominant gene $+w_2$ is translocated to the w chromosome, so this gene is inherited by sex. Translocanted breeds are very silky-skinned, in the 80^x years of the last century components of hybrids were zoned for production and gave good cocoon yields. These are hybrids such as SANIISH 30 x Sov.-5,

C-5 x SANIISH 30, C-13 x C-14, C-14 x C-13 and others. Now breeds-components of these hybrids are in the living collection of mulberry silkworms of NIISH and are produced according to the group breeding scheme. In order for these breeds to be used

The table shows that, on average, the sex-labelled breeds over the 3 years, in hybridisation, selection and breeding work should be carried out with them to improve the main productive properties.

§ 3.2 Breeding and pedigree work with sex-labelled C-5 breed at the grena stage

In gren production, an important problem is the proper and timely separation of cocoons by sex to prepare hybrid gren without admixture of pure breeds.

Preparation of hybrid gren for industrial fattening, which is not contaminated with less productive gren of other breeds, requires accurate sex separation of breeding material in order to cross females of one breed with males of another breed.

Sex-labelled breeds at the egg stage solve the problem of preparing hybrids that are not contaminated by less productive initial forms.

Figure 3.2.1 shows the cocoons of the C-5 breed.

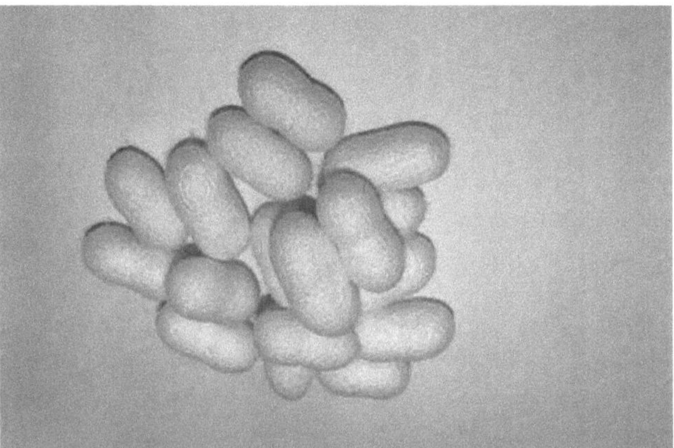

Fig. 3.2.1 Cocoons of C-5 breed

Eyes of C-5 female butterflies are dark and white in males. Silkiness of C-5 is 24-25%. In breeding works females and males feed together. The gren is incubated separately, in the second instar 110 females and 110 males are counted from each breeding clutch for feeding, 220 caterpillars in total. Each breeding clutch is divided by sex into dark and white grena. Normal, unfertilised and dried eggs are counted. All indicators are counted. The breeding stock is retained with a $1 : $1 ratio to maintain sex-labelling.

The caterpillars of the breed have a pronounced mask on the cover, the cocoons are elongated.

The C-5 breed in hybrid combination with a parthenogenetic clone once passed station trials and was submitted to state trials.

At present, the breeds are not reproduced at the silk breeding stations, so selection and breeding work with them is carried out at the NIISH in order to maintain, breed and further

improvements in a number of productivity traits.

To illustrate the results, Figure 3.2.2 shows the silkiness of C-5 cocoons by year of selection.

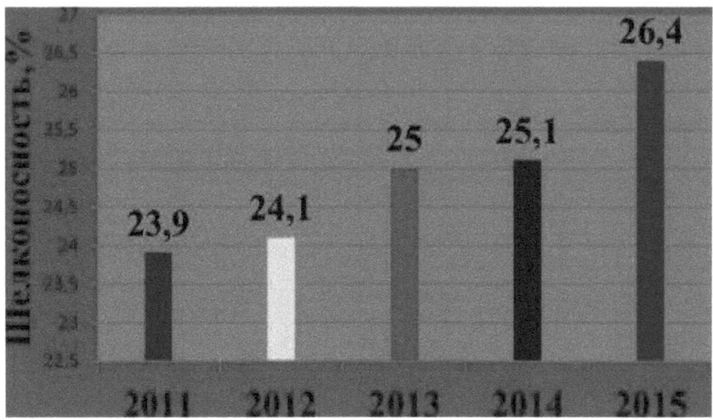

Fig. 3.2.2 Silkiness of C-5 cocoons by years

Figure 3.2.2 clearly shows the dynamics of silkiness growth by years of selection. The increase in silkiness from 23.9% in 2011 to 26.4% in 2015 indicates a high genetic potential of the breed and the possibility to successfully use C-5 in hybridisation.

Biological parameters of C-5 breed on all fed families by breeding years are presented in Table 3.2.1.

The gregariousness of all incubated families of Sovetskaya 5 breeds was increased from 92.8 to 95.0%. The gregariousness of families allowed for fattening was 97.4%.

The breed has an inherent cocoon mass, on average equal to 1.86 g, covaried by year of rearing by silkiness. Low cocoon mass was recorded in 2012. Viability of caterpillars increased insignificantly and did not reach the planned level. The silkiness of cocoons increased from 23.9 to 26.4% according to the conducted selections.

Biometric processing of indicators revealed that such biological traits as caterpillar viability and cocoons weight vary to the greatest extent in families of C-5 breed. In terms of silk content in raw cocoons, the breed is relatively equalised (Table 3.2.2).

Table 3.2.1.

Biological indicators of the C-5 breed (2011-2015)

Years	Number of families selected for incubation	Gene revival of incubated families, %	Number of families selected for fodder U	Revitalised families, % M±t	Caterpillar viability, % M±t	Weight		Silkworms - cocoon NOS, % M±t	Butterfly emergence, %	Number of grouse, %
						Cocona, g M±t	Silk shell, mg M±t			
2011	158	92,8±0,42	60	97,5±0,45	75,5±0,89	1,76±0,013	421±0,15	23,9±0,013	-	9,38
2012	105	91,14±0,54	43	96,6±0,18	87,9±0,15	1,64±0,017	395±5,0	24,1±0,15	10,0	2,88
2013	133	94,7±0,	18	98,2±0,	89,1±1,	1,82±0,	457±8,	25,0±0,	6,7	6,62

3		28		19	39	028	6	11		
201 4	103	97,0±0, 41	18	98,7±1, 18	91,2±1, 5	2,11±0, 14	528±0, 18	25,1±0, 2	7,2	8,6
201 5	71	95,0±0, 58	51	96,1±0, 28	78,4±1, 2	2,06±0, 5	550±3, 41	26,4±0, 48	8,4	10,71
Cf.	570	94,1±0, 57	190	97,4±0, 42	81,2±1, 02	1,86±0, 13	470±4, 06	24,9±0, 19	8,0	7,63

Table 3.2.2.

Variation of families of the breed Sovetskaya 5 by biological indices for 5 years

By year	Coefficient of variation, %			
	The revitalisation of the grena	Caterpillar viability	Cocoon weight	Percentage of silk shell
2011	3,5	8,87	5,7	3,64
2012	1,3	7,89	7,36	3,95
2013	0,81	7,09	6,59	1,87
2014	0,68	7,9	2,9	3,40
2015	0,81	8,63	3,57	1,87
Average	1,42	8,0	5,2	2,94

As can be seen from Table 3.2.2, the coefficients of variation of biological parameters of the breed are not high, i.e. the C-5 breed is rather stabilised. The data of Table 3.2.3 show that females intended for the breeding group were selected with intensity equal to 61.1% on average.

Table 3.2.3.

Intensity of selection of breeding families of the C-5 breed and their five-year track record

Year	Number of families taken in		Selection intensity, %	Gene revitalisation, %	Caterpillar viability, %	Cocoon weight, g	Weight of silk shell, mg	Silk sheathing KH, %
	Youkor m-ku	tribe						
2011	60	32	53,3	98,0	81,8	1,75	420	24,0
2012	43	27	62.6	96,7	90,6	1,68	404	24,0
2013	18gr.	18gr.		98,3	83,7	1,82	463	25,4
2014	18gr.	4gr.		99,2	86,5	2,16	524	25,4
2015	51	35	68,6	96,0	81,8	2,08	551	26,5
Medium	190	114	61,1	97,6	84,8	1,90	472,4	25,06

Table 3.2.4.

Breeding differential by biological indicators of C-5 breeding families over five

years

Years	Revitalisation (%) of family clusters			Viability (%) of caterpillars of families			Cocoon weight (g) families			Weight (mg) of silk sheath families			Content (%) of silk sheath families		
	take incubation	pedigrees	S	take incubation	pedigrees	S	take incubated	pedigrees	S	take incubated	pedigrees	S	take incubated	pedigrees	S
2011	92,8	98,0	5,25	75,5	81,8	6,5	1,76	1,75	0,01	421	420	1,0	23,9	24,0	0,1
2012	91,14	96,7	5,56	87,9	90,6	2,7	1,04	1,68	0,04	595	404	9,0	24,1	24,0	0,01
2013	94,7	98,5	5,6	85,1	85,7	0,6	1,82	1,82	0,0	457	463	6,0	25,1	25,4	0,3
2014	96,7	99,5	2,8	89,4	95,4	6,0	2,09	2,09	0,0	528	555	2,0	25,4	25,7	0,3
2015	95,0	96,0	1,0	78,4	81,8	3,4	2,08	2,08	0,0	550	551	27	26,5	26,4	0,1
Middle.	94,0	97,7	3,78	82,8	86,6	3,8	1,87	1,88	0,01	4,70	478,6	8,6	25,0	25,1	0,16

The data of Table 3.2.4 show that the highest values of selection differentials are characteristic of such biological traits as grena revival and viability. The difference in grena liveness between all incubated families and those selected for breeding averaged 3.7, and in viability - 3.8%.

The presence of selection differential made it possible to select for breeding. families with 97.7 per cent gene revival and 86.8 per cent viability.

In terms of cocoon and sheath weight, percentage of silk sheath, it was not possible to select breeding families differing significantly from the whole set of families. For these indicators, especially for silkiness, there was a difference in breeding individuals selected by individual analysis. The data of Table 3.2.5 show that breeding individuals, from which clutches of initial material were prepared, had a selection differential by cocoon weight on average for 5 years 0.04 g, silk sheath weight 20 mg, silk content in the cocoon 0.9%.

Table 3.2.5.

Breeding differential on biological indicators of breeding individuals of the C-5 breed over five years

By year	Cocoon weight, g.			Weight of silk shell, mg			Percentage of silk shell		
	total material	breeding stock	S	total material	breeding stock	S	total material	breeding stock	S
2011	1,76	1,78	0,02	421	420	-1	23,9	25,5	1,6
2012	1,84	1,71	0,07	395	404	9	24,1	24,7	0,6
2013	1,82	1,86	0,04	437	480	23	25,1	25,8	0,7

2014	2,10	2,13	0,03	525	555	27	25,1	26,2	1,0
2015	2,08	2,08	0	550	975	23	28,4	27,0	0,6
Middle.	1,88	1,91	0,04	470	486	20	24,9	25,8	0,9

Thus, the breeding individuals were characterised by cocoon weight 1.91 g, silk content in cocoon 25.8%, quite meeting the planned parameters.

Technological properties of the breed's cocoons were studied by analysing a sample taken from all families (Table 3.2.6).

Table 3.2.6.

Technological indicators of C-5 rock

By year	Weight of dry cocoon, g	Raw silk yield, %	Total silk products, %	Metric yarn number, %	DNRH, m	Unwindability of the shell, %	Production length of yarn, m
2011	0,747	46,9	51,82	3058	757	90,5	1011
2012	0,706	44,58	51,27	3159	776	88,24	996
2013	-	-	-	-	-	-	-
2014							-
2015	0,984	42,74	54,51	2547	821	78,55	1183
Middle.	0,812	44,7	52,50	2914	784	85,00	1063

As can be seen from Table 3.2.6, the technological indicators of C-5 did not change by years.

§ 3.3 Breeding and pedigree work with sex-labelled C-10 breed at the grena stage

The C-10 breed is labelled by sex at the grena stage (grena dark - $$, light brown - $$). Breeding work in the genetics department was started in 2013. The breed was bred in families.

Shape of cocoons C-10 elongated with slight interception, with fine granularity. Female butterflies lay dark and light brown grenae. From the dark ones $$ is inoculated, from the light brown ones $$ is inoculated. The $$ has dark eyes and the $$ has light brown eyes. Vitalisation of grena is 96,8% on average, viability of caterpillars is 89,90%, butterfly emergence is 7-8%, cocoon unwinding rate is 88-89%, metric number is 3000-3100 units. The caterpillars go to curling very friendly.

Table 3.3.1.

Biological indicators of C-10 breed for three years

Years	Number of seeds, selected for incubation, pcs.	Revitalisation of the stock exchange. families,%	Number of families, selected for uprooti	Fattening, families, % M±t	Caterpillar life M±t	Weight		Co. silk wrap, %.	Number of deaf %	Butterfly emergence, %
						cocoon, g.	Shells, mg			

			ng, pcs.							
2013	111	95,6	66	90,2±0,2	90,2±0,6	1,58±0,01	414±0,2	26,2±0,3	3,3	5,5
2014	100	94,0	55	97,4±0,6	90,5±0,5	1,62±0,01	426±4,45	26,3±0,1	2,6	6,6
2015	40	98,5	45	96,5±0,6	81,6±1,1	1,72±0,01	430±2,9	25,0±0,6	2,0	10,89
Cf.	87	96,03	55	97,3±0,4	87,3±0,7	1,64±0,01	423±2,6	25,8±0,2	2,0	7,76

On average, the gren revival of the incubated families is 96.3 per cent of the selected families - 98.5 per cent. This figure is not bad and a silkiness of 25.8% is also considered high.

To illustrate the results in Figure 3.3.1, we present cocoon weight indices of C-10 breed for 3 years of sampling.

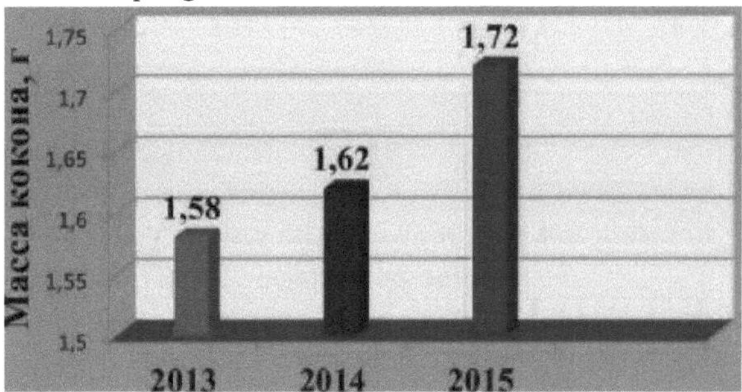

Fig. 3.3.1. Cocoon weight of C-10 breed by year of selection

Figure 2.3.1 shows how the cocoon weight of C-10 breed grows during 3 years of sampling. If in 2013 the cocoon weight was 1.58 g, in 2015 its value reached 1.72 g. At the same time, the silkiness of the breed was 26.2%, 26.3%, 25.0% - very high values.

Biometric processing of the data showed that families of the C-10 breed differed in the main indices and were equalised in cocoon weight (Table 3.3.2).

Table 3.3.2.

Coefficient of variation of families in C-10 breed by biological traits for three years

By year	Revitalising grenades %	Life mode. gus.,%	Cocoon weight,g	Silk content %
2013	2,0	5,0	6,3	2,3
2014	4,6	6,8	6,2	3,4
2015	1,0	8,2	5,8	8,2

| Middle. | 2,5 | 6,66 | 6,1 | 4,6 |

The caterpillars varied in viability 6.66% and silk content 4.6% (Table 3.3.2).

Table 3.3.3 shows the number of families sampled and their characterisation.

Table 3.3.2.

Intensity of selection of breeding families and their biological parameters in C-10 breed

By year	Number of families taken			Inten. Selection of families	Revitalised. Greny %	Caterpillar viability, %	Weight		Content. Masonry, %
	Inc.	Youkor -	Breeding -				Koko and g.	silk. Obol., mg.	
2013	129	66	38	57,6	95,6	91,1	1,58	414	26,2
2014	105	55	35	63,6	97,5	89,0	1,59	405	25,5
2015	45	45	36	80,0	98,5	81,7	1,72	430	25,0
Total	279	166	109	67,06	97,2	87,2	1,63	416	25,5

From the results, females destined to prepare clutches of source material were selected at an average intensity of 67.06%.

The intensity of selection was judged by selection differential, calculated by the excess of breeding families' performance over the performance of incubated families. The value of selection differential is given in Table 3.3.4.

Table 3.3.4.

Selection differential (S) on biological indicators of breeding families of C-10 breed

Live births, % of families		S	Gisnespo. Gus. % of families		S	Cocoon weight,g families		S	Mass of shells, mg families		S	Silk content, % of families		S
proincubus	tribes.		Vyck.	pl.		Vyck.	breeding		youk.	breeding		youk.	breeding	
90,0	95,6	5,6	91,1	93.1	2,0	1,58	1,56	0,02	414	410	4	26,2	26,3	0,1
94,0	97,4	31	89,0	90,5	1,5	1,61	1,52	0,01	410	426	16	25,5	26,3	0,8
98,5	98,5	0	79,6	81,7	2,1	1,72	1,72	0	421	430	9	24,6	25,0	0,4
94,1	97,1	3,0	80,5	88,4	1,9	1,63	1,63	0	415	432	7	25,4	25,83	0,43

From the data of Table 3.3.4, it can be seen that the highest value of selection differential is observed for grena revival and shell weight. In terms of cocoon weight, percentage of silk sheath it was not possible to select breeding families significantly different from the totality of families.

Characteristics of breeding individuals of the breed are given in Table 3.3.5.

Table 3.3.5.

**The value of selection differential (S) by biological
the performance of C-10 breeding individuals over three years**

Years	No. of proa- nal IND. Cocoons, pcs	Select breeding. Speciality. pcs	Cocoon weight,g		S	Shell weight, mg		S	silk, %		S
			Total	breeding individuals		Total	breeding individuals		total material	breeding. Specimens	
2.13	918	694	1,56	1,56	0	396	404	8,0	24,8	29,9	1,1
2014	1068	621	1,61	1,62	0,01	410	425	16,0	23,5	26,3	0,8
2015	1812	662	1,84	1,74	0	440	444	4	25,3	25,5	0,2
Cf.	939	659	1,63	1,64	0,01	415	424	9,0	25,2	27,2	2,0

The results show that the breeding individuals selected for preparation of grena had: cocoon mass 1.04 g, silk sheath mass 424 mg, silk content in raw cocoon 27.2%.
Technological properties of cocoons were determined by unwinding a common sample. Their results are given in Table 3.3.6.

Table 3.3.6.

Technological indicators of cocoons of C-10 breed for three years

Years	Weight of dry cocoon, gr-	Total, %		Metric yarn number,units.	DNRH,m	Dividend of the total, %	Production yarn length,m.
		raw silk	total silk product.				
2013	0,932	34,75	40,84	2881	791	84,98	990
2014	0,729	43,0	54,30	3152	764	79,42	987
2015	0,750	40,15	50,15	2901	581	80,85	888
Cf.	0,803	39,30	48,43	2979	712	81,75	955

On average, for three years the breed is characterised by raw silk yield of 39.3%. It has a fine thread of 2979 units and
production length of 955 metres of yarn.
Figure 3.3.2 shows the cocoons of the C-10 breed.

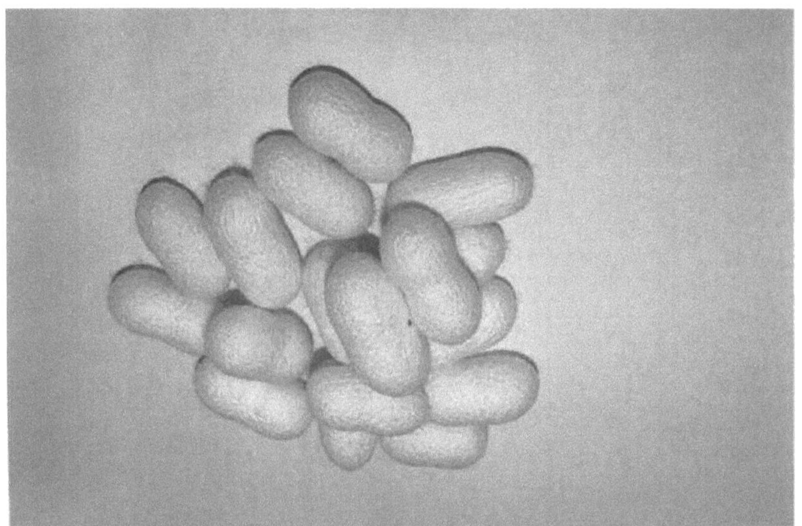

Fig. 3.3.2 Cocoons of C-10 breed

Eyes of females of breed C-10 are dark, of males - light brown, cocoons are elongated, with an intercept, silkiness in raw cocoons - 24-25%, cocoon unwinding - 88-89%, length of continuous unwinding - 900-1000 metres, production length of thread - 1200-1250 metres. Grena should be divided into dark and light brown. A sex ratio of $1:1^ must be maintained annually, otherwise this quality can be lost quickly. All sex-labelled breeds are highly silky.

At this time, this breed is conserved in the mulberry silkworm gene pool.

§ 3.4 Breeding and pedigree work with sex-labelled C-12 breed at the grena stage

Breeding work was carried out for 5 years in the spring on seedling rearing to improve grena revival rates, caterpillar viability and retention of high silk in the cocoon.

From the data in Table 2.4.1, it can be seen that the breed has a high grena revival rate of 97.1%, while for all incubated families it is 92.2-96.2%. Viability of caterpillars, on average, is equal to 80.6%, silkiness of cocoons is increased from 24.0 to 26.2%.

The families of the breed vary to the greatest extent in caterpillar viability (8.92%) and cocoon weight (5.80%). In terms of silk content in raw cocoons, the breed is significantly equalised (4.30) (Table 3.4.2).

Table 2.4.3 shows that the intensity of selection of breeding families from all fed families averaged 55.48%. The selection intensity was judged by the value of selection differential of biological parameters of breeding families.

Table 3.4.1.

Biological indicators of C-12 breed for five years

Yea	Number	Gene	Numb	Revitalisat	Caterpilla	Weight		Cocoon	Numb

rs	of seeds selected for incubati on	revival of incubate d families, %	er of famili es sent for fatteni ng	ion of grens of fattened families, % M±sh	r life, % M±sh	cocoon, g M±sh	silk shell, mg M±sh	silkiness, % M±sh	er of grous e, %
201 1	144	96,2±0,2 5	56	96,5=1=0, 13	74,6±1,17	1,61±0,0 14	387±3,7 2	24,0±0,01	12,57
201 2	139	94,0±0,3 2	45	97,1=1=0, 75	87,5±0,74	1,62±0,0 1	382±2,6	23,6±0,01	5,55
201 3	103	94,8±0,4 1	56	97,4±0,22	83,5=1=0 ,98	1,78±0,0 1	445±3,7	25,0=1=0 ,01	5,57
201 4	148	92,6±0,3 6	51	97,6=1=0, 18	82,8±1,5	1,81±0,0 1	474±4,2	26,2±0,18	7,26
201 5	PO	92,2±0,6 0	61	97,0±0,4	75,0±1,03	1,89±0,0 15	494±4,1 7	26,0±0,11	12,4
Cf.	644	93,96±0, 38	269	97,12=1=0 ,33	80,68±1,0 8	1,74±0,0 1	436,4±3, 67	24,9±0,08	8,67

Table 3.4.2.

Variation of families of C-12 breed by biological indices by years

Years	Coefficient of variation (Cv)			
	The revitalisation of the grena	caterpillar viability	cocoon weight	% silk sheath
		10,4	6,80	3,40
2011	7,74	10,1	4,32	4,20
2012	5,25	8,0	5,06	6,20
2013	1,6	7,25	6,5	4.98
2014	1,33	8,85	6,34	2,73
2015	3.2	8,92	5,80	4,30
Cf.	3,76			

Table 3.4.3.

Intensity of selection of breeding families of C-12 breed and their characteristics by years

Number of families taken in			Intensity of selection of breeding families from the number of fattened families, %	Grena revitalisation,%	Caterpillar viability, %	Weight		% silk shell	% deaf arey
incubation	chicken fattening	tribe				cocoon,g	silk shell, mg		
144	56	32	57,0	98,3	81,0	1,62	393,8	24,3	8,72

139	45	24	53,3	97,2	91,4	1,64	398,0	24,2	4,25
103	56	28	50,0	97,5	90,2	1,79	450,0	25,1	5,03
148	51	38	74,5	97,5	84,4	1,81	474,0	26,5	6,53
PO	61	26	42,6	95,0	82,6	1,91	503,0	26,3	8,4
644	269	148	55,48	97,1	85,92	1,75	443,7	25,2	6,58

Table 3.4.4.

Breeding differential on biological indicators of breeding families of C-12 breed

Gohholes	Grena revival, % of families		S	Viability. gus. %, families		S	Cocoon weight,g families		S	Weight of silk, % of families		S	% silk sheath, families		S
	incubated	pedigrees		taken for fattening	Tribal		fattened	Tribal		taken for fattening	plume of		taken for fattening	pedigrees	
2011	96,2	98,3	2,1	74,6	81,8	6,4	1,61	1,62	0,01	387	393	6	24,0	24,3	0,3
2012	94,0	97,1	3,1	87,5	91,4	3,9	1,62	1,64	0,02	382	398	16	23,6	24,2	0,06
2013	94,8	97,5	2,7	83,5	90,2	6,7	1,78	1,79	0,01	445	450	5	25,0	25,1	0,1
2014	92,6	97,5	4,9	82,8	84,4	1,6	1,81	1,81	0,00	474	474	0	26,2	26,5	0,3
2015	92,2	95,0	2,8	75,0	82,6	7,6	1,89	1,91	0,02	494	503	9	26,0	26,3	0,3
Cf.	94,0	97,0	3,0	80,7	86,0	5,3	1,74	1,75	0,01	436	443,6	9	24,9	25,3	0,2

Table 3.4.5.

Characterisation by biological indicators of breeding individuals of the C-12 breed by years

Years	Cocoon weight, g			Weight of silk sheath, %			% silk sheath		
	total material	breeding stock	S	total material	breeding stock	S	total material	breeding stock	S
2011	1,61	1,63	0,02	387	412	25	24,0	25,3	1,3
2012	1,62	1,64	0,02	382	411	29	23,6	25,1	1,5
2013	1,78	1,79	0,01	445	460	15	25,0	25,7	0,7
2014	1,81	1,83	0,02	474	486	12	26,2	26,6	0,4
2015	1,89	1,96	0,07	494	531	37	26,0	27,1	1,1
Cf.	1,74	1,77	0,03	436	460	24	24,9	26,0	1,1

The data of Table 3.4.4 show that the highest values were grena revival and caterpillar viability. At the same time, the gap in grena revival between all incubated families and those selected for breeding was 3.0 and in viability 5.3%. As a result, grena revival of breeding families was 97.0%, caterpillar viability 86.0%.

It was not possible to select breeding families that differed significantly from the whole population of families in cocoon and sheath weight, percentage of silk sheath. Breeding individuals were selected by individual selection method.

Data in Table 3.4.5 show that the breeding individuals selected for initial clutches had cocoon weight of 1.77 g, silk sheath of 460 mg, silk content in the raw cocoon - 26.0%. Breeding individuals in the 2015 season were particularly characterised by high performance, silkiness, for example, was 26.6-27.1%.

To illustrate the results obtained, Figure 2.4.1 shows the values of silk sheath weight of the C-12 breed by year of selection.

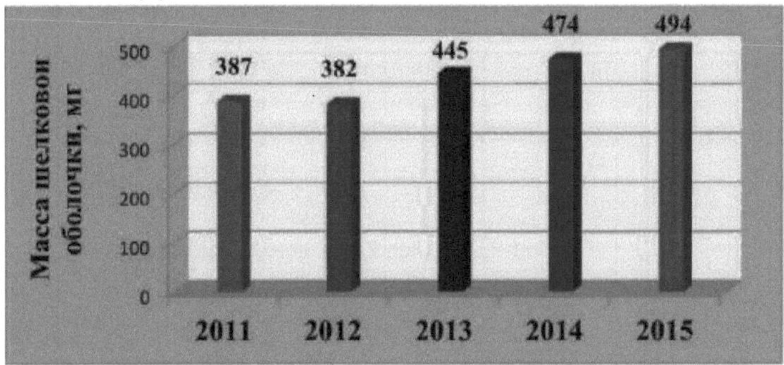

Fig. 3.4.1. Weight of silk sheath of C-12 breed by year of selection

Figure 3.4.1 clearly shows how 5 years of selective breeding has raised the cocoon mass of the C-12 breed from 387 mg in 2011 to 494 mg in the 2015. The cocoon weight and silkiness increased simultaneously with the shell weight (Table 3.4.1).

Technological properties of the breed's cocoons were determined by unwinding the total image made up of all reared families. On average, according to five-year data, the breed is characterised by a raw silk yield of 43.37%, a thread fineness of 3156ed. and a production thread length of 1150m.

Table 3.4.6.

Technological indicators of C-12 breed by years

Years	Weight of dry cocoon, gr-	Total, %		Metric yarn number,units.	DNRH,m	Unwindability of the shell, %	Production yarn length,m.
		raw silk	of all silk products.				
2011	0,661	45,15	51,17	3531	705	88,2	1105
2012	0,712	42,44	50,24	3269	743	84,64	1003
2013	0,905	44,25	52,06	3020	841	85,01	1283
2014	0,875	43,3	50,89	3062	822	85,1	1184

| 2015 | 0,900 | 41,25 | 90,43 | 2901 | 751 | 81,75 | 1177 |
| Cf. | 0,810 | 43,27 | 90,95 | 3156 | 772 | 84,94 | 1150 |

Figure 3.4.2 shows the cocoons of the C-12 breed.

Fig. 3.4.2 Cocoons of C-12 breed

Cocoons of breed C-12 are oval-rounded, very silky - 2425%. Grenae are dark and white. From dark eggs females emerge, from light-coloured eggs males emerge.

The eyes of female butterflies are dark. The eyes of males are white.

Vitalisation of the gren is 96.6%, viability is 89.91%, and they go to curling in a friendly manner.

§ 3.5 Use of parthenoclones as a parent breed for industrial hybridisation

Isogenic female parthenoclones can be used as a mother breed for interbreed hybridisation. B.L. Astaurov wrote about this possibility [24;p.130-141]. The advantages of using parthenoclones for interbreed hybridisation are as follows.

1. The F1 hybrid, obtained from crossing parthenoclon females with males of the conventional breed, differs from conventional hybrids by greater hardiness and equalisation in all traits.

2. Genetic constancy of parthenoclon females will allow to exclude labour-intensive, expensive breeding work in scientific-research institutions and at breeding silk-breeding stations.

3. The presence of single females in parthenoclones is also useful for production, as it will allow the preparation of hybrid grens of 100% purity, without resorting to labour-intensive, expensive and very inaccurate separation of females from males at the cocoon stage for the purpose of interbreeding.

The economic effect of preparing hybrid grena with parthenogenetic females will be a large amount.

Hybrids from crossing parthenogenetic females with normal males of other breeds are characterised by high qualities (Table 3.5.1).

Table 3.5.1.

Characteristics of mulberry silkworm hybrids obtained from crossing females of parthenoclones of SANIISH 30 breed with males of C-5 breed (2011)

Hybrid	Paul	G of inability,%			Weight		Silkiness, % '	Silk yield, %
		embryo nal	postamb rnonal nal	KOKO pa г		shell,mg		
Tetragnbride 3 (control)		97,0	98 Л	2,15		482	22,4	100,0
PC 5M0xC- 5		98,7	95,6	2,15		533	24,8	112,7
PC 153x05	'd	99,0	982	2,08	301		25.2	117,7

From the data given in Table 3.5.1, obtained under good ecological conditions of caterpillar feeding, it follows that the new hybrids exceed the control hybrid Tetrahybrid 3 in viability, silkiness, raw silk yield and other indicators. Under more severe conditions, this difference will be even greater.

Under production conditions, this hybrid had a yield of 74 kg of cocoons per box of grena against 64.5 kg of the control hybrid Tetrahybrid 3. For the first time in the history of world silk breeding, production trials proved the possibility of hybrid grena preparation from parthenoclones and high productivity of hybrids obtained from them under production conditions.

As a result of station trials, the best hybrid combinations involving parthenoclones were singled out. These hybrids were in due course accepted for testing by the State Crop Testing Commission.

Table 3.5.2 shows the biological parameters of the created clones of mulberry silkworm.

Table 3.5.2.

Biological parameters of parthenogenetic clones (2011-2013)

Name of clones	Parthenogenes gus. yield,%	Caterpillar viability, %	Cocoon weight, g	Weight of shell, mg	Silk content, %
A-153 pc	75,33±0,33	89,1±1,91	1,90±0,056	402±13,05	21,2±0,12
A-218pk	87,0±0,29	82,7± 1,21	1,90±0,042	394±9,54	20,6±0,175
A-238 pk	89,5±0,65	82,7±0,87	1,80±0,018	368±1,76	20,4±0,196
A-261 pc	87,67±2,41	90,0± 1,18	1,91±0,09	370±14,4	19,4±0,250
51 -40 pk.	86,02±0,86	89,1±1,04	1,86±0,07	368±16,38	19,8±0,29
9-pk	80,0±1,92	88,5±1,27	1,61±0,07	354±12,8	20,7±0,23
43-pk	85,4±0,83	85,9±1,88	1.8 BUT.06	355±31,4	21,8±0,32
29-pk	91,33±77	95,9±1,20	1,74±0,09	238±10,02	13,7±0,25
SANIISH-30 (k)	97,6±0,56	82,7±1,43	1,66±0,013	3,31±4,35	23,6±0,199

As follows from the table, the yield of parthenogenetic caterpillars varies between

50

75.33-91.33%, which is lower than in the control oboepole breed SANIISH-30. This may be explained by the fact that clones are represented by females and they are less productive sex. Cocoon weight varies from 1.61 g to 1.91 g, sheath 238-402 mg, silkiness from 13.7 to 21.8%. Special attention should be paid to clones AIC, 9pk, A-153pk, A-218pk, 51-40pk, which are widely used for hybridisation.

Figure 3.5.1 shows cocoons of parthenogenetic clones.

Figure 3.5.2 shows female parthenoclone butterflies.

Fig. 3.5.1 Cocoons of parthenogenetic female mulberry silkworms

Fig. 3.5.2 Constant homogeneous butterflies - clone females

As it was mentioned above, parthenogenetic clones are represented by female individuals and this affects the biological indices, since mulberry silkworm males are 15-16% silkier than females and their viability is higher.

Next, we present a table with coefficients of variation for biological traits (Table 3.5.3).

Table 3.5.3.

Coefficients of variation of biological traits parthenogenetic clones (201102013)

Name of clones	Caterpillar viability, %	Weight		Silkiness of cocoons, %
		cocoon,g	shell,mg	
A-153 pc	6,59	7,88	7,90	1,44
A-218pk	8,20	5,26	8,84	12,80
A-238 pk	5,2	5,38	12,52	0,38
A-261 pc	16,6	3,19	4,27	2,84
51.40 pk.	9,34	4,19	5,32	3,08
113 pk.	6,11	8,80	5,98	2,44

The data of the table show that variations in biological traits are not large, which confirms the homogeneity of parthenogenetic clones. The following table shows the technological analysis of parthenoclones (Table 3.5.4).

Table 3.5.4.

Technological indicators of parthenoclones (2011-2013)

Names. Clones	Weight of dry coke, g.	Output		Metric. Number, units.	DNRCN, m.	Unwind., %	Production. Length, m.
		raw silk,%	silko prod.				
A-153 pc	0,630	41,27	48,81	3110	6681	84,55	831
A-218pk	0,537	35,24	42,11	3827	571	83,69	735
A-238 pk	0,599	36,64	42,90	3541	595	85,88	804
A-261 pc	0,542	43,02	49,71	3663	627	86,54	864
51.40 pk.	0,521	37,04	46,06	3991	585	80,42	731
113 pk.	0,460	33,70	41,66	2859	576	82,88	914

Table 3.5.4 shows that the technological parameters of parthenoclones are quite high. Metric number of clone 5140pk reaches 3991 units, and A-218pk - 3827 units. Unwinding capacity of cocoons of clone A=238pk - 85,88%, clone A-261 - 86,54%. The silk yield of A-153pk was 48.81%, that of A-261pk - 49.7%.

Such high technological characteristics of cocoon thread of cocoons once again confirm the value of cocoons as components for hybridisation, as clones not only provide heterosis and stability of biological parameters, but also can change for the better the technological properties of silk thread.

Figure 3.5.3 shows cocoons of clones 113pc and 29pc, which may also be of interest to breeders. These are yellow-cocoon clones.

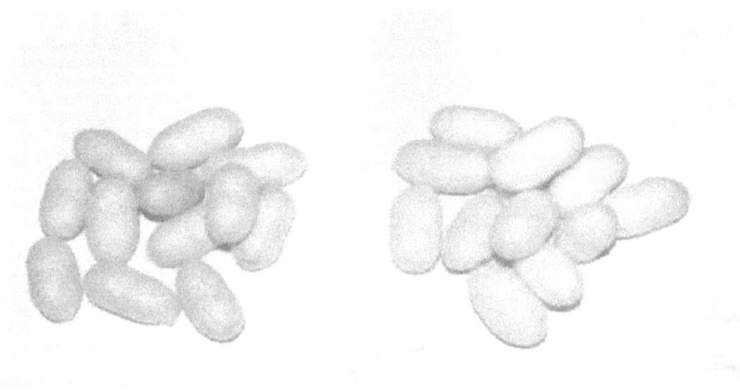

Fig. 3.5.3. Clone-113 cocoons on the left, Clone-29 cocoons on the right

Figure 3.5.3 shows the cocoons of Clone-29 and Clone-113. These clones were obtained more than 80 years ago. Viability of caterpillars is very high, silkiness 14.5-15.0%. They consist of one female. They are used for breeding works.

§ 3.6 Laboratory tests of hybrids in Nukus branch of TashGAU

Simultaneously with the improvement and maintenance of promising breeds and clones in this study, hybrids with the above-mentioned breeds and clones as their components have been created and tested for several years. Unfortunately, due to organisational and economic difficulties, breeds C-10, C-12 were not used in hybridisation. It should be noted that hybrids are tested in large numbers (21 hybrids). Below in Table 3.6.1 we give the characteristics of these hybrids.

It follows from the table that the best hybrids were: C-13 x C-14, Meechennaya 1 x Meechennaya 2, SANIIISH 30 x Meechennaya 1, C-5 x SANIIISH 30, 9 pk x Meechennaya 1, 9 pk x C-5. All these hybrids are characterised by good viability and silkiness.

Table 3.6.1.

Biological indicators of hybrids (2011-2013)

Name of hybrids	Life span. Caterpillars, %	Weight cocoon,g	shell, mg	Contained. Silk, %
C-13 x C-14	97,1	1,90	478	25,2
Sworn 1 x Sworn 2	98,0	2,04	487	23,8
SUNYISH 30 x Marhamat	97,6	2,14	499	23,3
Marhamat x SANIISH 30	96,7	2,01	447	22,2
SANIISH 30 x Swabbed 1	94,5	2,0	485	24,3
Swabbed 1 x SANIISH 30	95,1	1,76	423	24,0
C-5 x SANIISH 30	97,1	2,0	475	23,7
AF238 pk x Mechennaya 1	95,1	2,09	469	22,4

9 pk x Mechennaya 1	97,1	1,92	472	24,6
AF153 pk x Mechennaya 1	92,2	2,14	484	22,6
51.40 pk x Sword 2	95,2	1,85	432	23,3
A-261 pk x Mechennaya 1	96,0	1,87	419	22,4
A-218 pk x Mechennaya 1	92,4	1,63	372	22,8
A-238 pk x C-5	96,7	2.07	470	22,7
A-261 pk x C-5	98,0	2,09	446	21,3
9 pk x C-5	97,8	1,94	455	23,4
A-153 pk x C-14	95,3	2,15	481	22,3
A-238 pk x C-14	98,4	2,0	452	22,6
51.40 pk x C-14	98,9	2,02	473	23,3
9 pk x C-14	95,5	1,95	452	23,1
A-238 pk x Orzu	98,7	2,17	506	23,2

High viability of caterpillars of practically all investigated hybrids attracts attention. But the best hybrids in terms of viability are hybrids: Mechennaya 1 x Mechennaya 2 - 98.0%, A-261 PC x C-5 - 98.0%, A-238 PC x C-14 - 98.4%, 51.40PC x C-14 - 98.9%, A-238 PC x Orzu - 98.7%. Interestingly, one of the partners of these hybrids are parthenogenetic clones.

High cocoon weight (more than two grams) is shown by hybrids: Mechennaya 1 x Mechennaya 2, SANIISH 30 x Markhamat, A-238 PC x Mechennaya 1, A-153 PC x Mechennaya 1, A-238 PC x C-5, A-261 PC x C-5, A-153 PC x C14, A-238 PC x Orzu.

Silkiness of the studied hybrids is rather high from 21,3% to 25,2%, the most silky hybrids were: C-13 x C-14 - 25,2%, SANIISH 30 x Mechennaya 1 - 24,3%, Mechennaya 1 x SANIISH 30 - 24,0%, 9PK x Mechennaya 1 - 24,6%. It should be noted that one of the components of these hybrids are sex-labelled breeds or parthenoclones.

The participation of sex-labelled breeds in hybrids makes the task of separating the material by sex even easier, since the breeding plants receive parthenoclon cocoons consisting only of females and sex-labelled cocoons consisting only of males. When sorting clonal cocoons for breeding, the majority of cocoons (86-88%) remain for breeding, as these cocoons are genetically homogeneous. Cocoons of males of sex-labelled breeds are sorted according to the existing rules. Hybrids between clones and sex-labelled breeds are easy to prepare and have 100% purity [113], [111;p.89], [112;p.343-349].

In a separate research experiment, hybrids of clones AIC, 9PK with breeds C-14, MCH, Ya-120 were tested. The results are given in Tables 3.6.2 and 3.6.3.

In 2015, 2016, 2017, all hybrids were fed with mixtures in three replications of 200 caterpillars each. Clutches with the best reproductive performance were selected for feeding (Table 3.6.2).

Table 3.6.2.
Reproductive performance of the hybrids under study by years

No. of items	Name of hybrids	years	Number of normal eggs, pcs	Weight of normal eggs, MG	Weight of 1 egg, mg
1	AIC x MG	2015	565	-	-
		2016	589	287	0,488
		2017	626	302	0,483
2	AIC x Ya-120	2015	594	-	-
		2016	595	257	0,472
		2017	639	323	0,506
3	AIC x C-14	2015	582	-	-
		2016	589	290	0,493
		2017	637	308	0,483
4	9PC x MG	2015	577	-	-
		2016	566	253	0,473
		2017	634	320	0,505
5	9PKxYa-120	2015	540	-	-
		2016	500	231	0,472
		2017	637	319	0,501
6	9PKxS-14	2015	572	-	-
		2016	570	221	0,470
		2017	622	304	0,488

Table 3.6.2 shows that the best in reproductive performance for all 3 years of research were hybrids AIC x C-14. AIC x Ya-120, AIC x MG. It is interesting that components of 2 of these hybrids are sex-labelled breeds C-14 and MG, which have changes in genomes.

For clarity, the data on the number of normal eggs in clutches of hybrids before and after selection are shown in Figure 3.6.1.

Figure 3.6.1 clearly shows how the number of eggs in clutches of clonal-breed hybrids changed after 3 years of selection. The hybrids AIC x Ya-120 (639 pcs), AIC x C-14 (637 pcs), 9PK x Ya-120 (637 pcs) were the best in terms of the number of normal grena in the clutch. The increase in the number of eggs in clutches of hybrids occurred in accordance with the increase in the number of eggs in clutches of breeds.

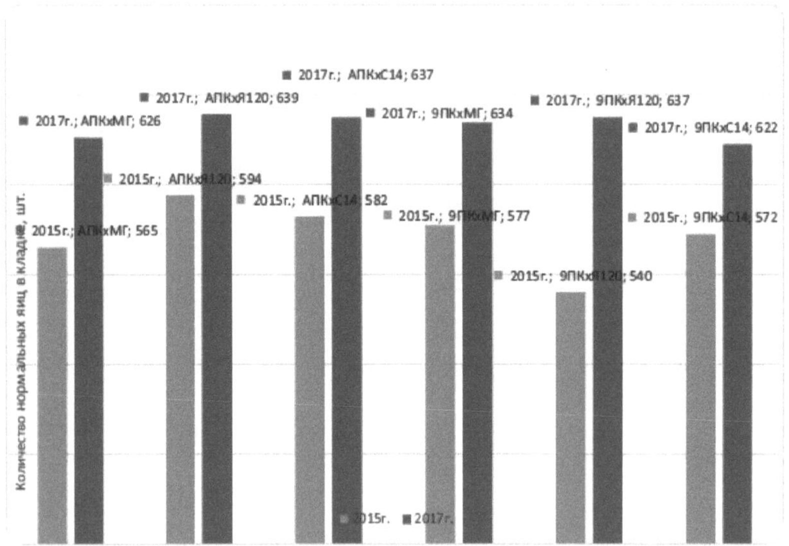

Figure 3.6.1 Number of normal eggs in clutches of hybrids before and after selection

Table 3.6.3.

Biological parameters of clonal-breeding hybrids by years.

№№	Hybrids	years	Caterpillar viability, %	Average weight		Silkiness, %
				cocoon, g	shell, mg	
1	2	3	4	5	6	7
1	AIC x MG	2015	87,2	1,67	379	22,4
		2016	96,6	1,87	424	22,8
		2017	98,0	1,45	322	22,3
2	AIC x Ya-120	2015	95,0	1,63	358	22,0
		2016	94,8	1,69	408	21,6
		2017	93,5	1,44	318	22,1
3	AIC x C-14	2015	95,7	1,62	385	23,8
		2016	94,0	1,85	407	22,2
		2017	94,5	1,65	369	22,3
4	9PC x MG	2015	94,3	1,54	341	22,1
		2016	97,2	1,91	421	22,0
		2017	93,7	1,73	370	21,7
1	2	3	4	5	6	7
5	9PKxYa-120	2015	93,7	1,65	369	21,4
		2016	97,1	1,96	418	21,3
		2017	93,7	1,65	382	23,2
6	9PKxS-14	2015	94,0	1,60	373	23,3

| | | 2016 | 95,5 | 1,84 | 394 | 21,6 |
| | | 2017 | 97,3 | 1,65 | 363 | 22,4 |

As can be seen from Table 3.6.3, viability was quite high (from 87.2 to 98.0%) during all years of the study. The caterpillars developed well, evenly, moulted in a friendly manner and quickly emerged on cocoons. Such behaviour is typical for caterpillars of clonal-breeding hybrids, as indicated by Strunnikov V.A. et al. [110;p.3-324], [113] and is very valuable in practical silk breeding.

At first glance, the indicators do not seem very high. However, one should keep in mind the other advantages of clonal-breeding hybrids, which were pointed out in the Introduction, and above all, the purity of preparation.

In 2017, the best in terms of caterpillar viability were hybrids: AIC x MG - 98.0%, 9PK x C-14 - 97.3%, by cocoon weight 9PK x MG - 1.73 g, AIC x C-14 - 1.65 g, by silkiness of cocoons 9PK x Ya-120 - 23.2%, 9PK x C-14 - 22.4%. According to the totality of the best biological indicators, we can note the hybrids: 9PK x C-14, AIC x MG. It is noteworthy that one of the components of the marked hybrids are sex-labelled C-14 and MG.

Breed C-14 is a genetically modified breed, but it has consistently high biological parameters with low coefficients of variation (Table 2.2.4), which indicates its genetic balance. The breed C-14 has been widely zoned throughout Uzbekistan since 1989 as a component of hybrids C-14 x C-13, C-13 x C-14 (certificate of invention No. 9003649 and No. 9003630). Hybrids with its participation showed excellent results.

The MG breed is sex-determined at the caterpillar stage, i.e. has a translocation in the genome, but despite this it has high performance (Table 2.2.4) and as a component of hybrids Meechenna 1 x Meechenna 2, Meechenna 2 x Meechenna 1 was zoned throughout Uzbekistan.

For clarity, we have shown the variation in silkiness level of hybrids by year in Figure 3.6.2.

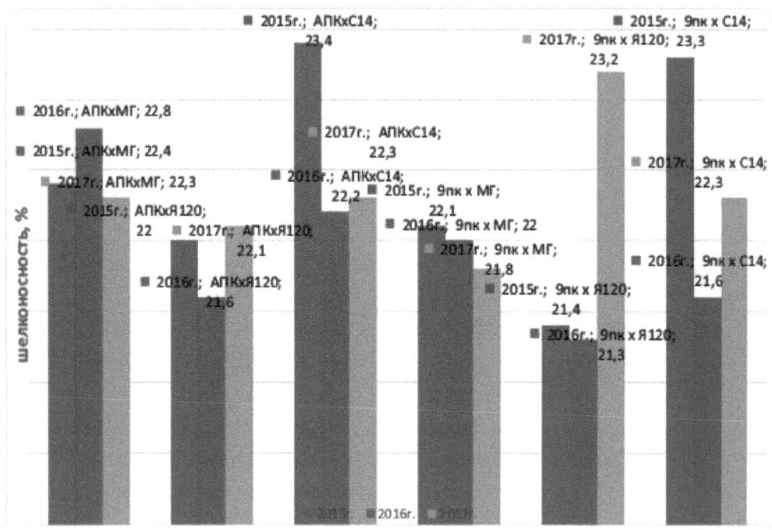

Fig. 3.6.2 Silkiness of hybrids by year

Figure 3.6.2 shows that the best 3 years for silkiness are hybrids AIC x MG (22.3 - 22.8%), AIC x C-14 (22.2 x 23.8%), 9PK x C-14 (21.5-23.3%).

The decrease in silkiness of hybrids in 2016 was in line with the decrease in silkiness of the component breeds. Early spring frosts in 2015, delayed start of forking, and feeding with immature leaves in hot weather led to a decline in many breed and hybrid performance in 2016. However, the early rainy spring of 2016 and good climatic conditions in 2017 favoured the optimal manifestation of the genetic potential of breeds and hybrids. The silkiness increased noticeably.

The technological performance of clonal-breeding hybrids was also collected.

Table 3.6.4.

Technological indicators of clonal-breeding hybrids by years

№	Name of hybrids	Years	Weight of 1 dry cocoon, g	Yield, %		Metric number of the thread	DNRCN, m	Total thread length, m
				raw silk, %	silk products			
1	AIC x C-14	2015	0,689	46,96	50,90	3880	648	654
		2016	0,702	44,93	48,79	3448	1067	1067
		2017	0,686	41,74	47,01	3436	942	1042
2	AIC x MG	2015	0,742	41,88	47,68	3636	908	1108
		2016	0,918	44,07	47,84	3333	800	1300
		2017	0,569	44,30	49,07	4065	883	1133
3	AIC x Ya-120	2015	0,610	47,33	50,10	3448	1050	1275
		2016	0,867	46,21	48,22	2994	1058	1158
		2017	0,629	42,92	48,11	3906	725	1183

		2015	0,682	45,26	50,01	3649	858	1275
4	9PKxS 14	2016	0,802	41,72	46,31	3300	775	1150
		2017	0,713	44,10	49,22	3356	867	1108
		2015	0,754	44,19	47,84	3558	1100	1150
5	9PC x MG	2016	0,772	42,66	48,50	3257	833	1150
		2017	0,708	41,12	47,50	3344	717	1150
		2015	0,700	42,48	46,62	3650	1108	1183
6	9PKxYa-120	2016	0,835	44,56	49,31	2958	1083	1108
		2017	0,708	45,56	50,36	3704	1008	1175

The data of Table 3.6.4 indicate that clonal-breeding hybrids are characterised by good technological properties. Attention is drawn to the high metric number of hybrid AIC x MG - 3333-4065 units, a long thread length of hybrid AIC x MG - 1108-1300 m. Such high technological indicators once again testify to the benefits of introducing clonal-breeding hybrids in industrial silk breeding in Uzbekistan.

The clonal-breeding hybrids underwent small-scale production trials for two years in Samarkand in silk breeding LLC "Mastura-Nur". According to silk growers' feedback, the tested hybrids were characterised by good viability and friendly development of caterpillars, had cocoons of approximately equal calibre.

Thus, for the first time in the history of silkworm breeding, clonal-breed hybrids of mulberry silkworm can be introduced into industrial production. Separate studies on the use of mulberry silkworm parthenoclones were carried out in Ukraine a few years ago, but they were not implemented.

Before purebreds were transferred to production, they were pre-fed in small station trials. The selected purebreds were then crossed for testing in small station trials.

Table 3.6.5.

Biological parameters of breeds used in hybridisation (2015)

No. No. of items	Name of breed	Caterpillar life, %	Gross weight		% silks, sheaths
			cocoon,g	shell, mg	
1	AGU-112	90,3±0,91	1,66±0,02	392±8,43	23,6±0,32
2	UzNIIISH-9	92,0± 1,23	1,75±0,03	41H9,20	23,5±0,23
3	Sworn 1	92,4±1,38	1,75±0,02	40HZ,54	23,3±0,15
4	Sworn 2	90,2± 1,64	1,680,04	383±8,17	23,0±0,24
5	SANIISH-30	89,9±1,82	1,72±0,01	387±12,53	20,7±0,71
6	C-5	91,3±2,62	1,73±0,09	392±6,49	22,9±1,05
7	C-5 clear goose.	88,B.O.B.,87.	1,54±0,06	32,3±7,46	21,1±0,54
8	C-10	88,9±1,65	1,56±0,05	362±7,69	23,5±1,24
9	C-12	88,4±1,49	1.7 BUT.08	408±8,25	24,2±1,46
10	C-13	91,2±1,72	1,57±0,02	398±20,16	25,4±1,23

| 11 | C-14 | 87,2±1,45 | 670±0,06 | 394±7,08 | 23,6± 1,04 |

All breeds have good caterpillar viability, as can be seen from Table 3.6.5. Breeds labelled by sex at the caterpillar stage are medium-skinned but highly silk-nosed: C-13 - 25.4%, C-12 - 24.2%, C-10 - 23.5%. The breeds labelled by sex at the caterpillar stage Mechennaya 1, Mechennaya 2 also have good silkiness 23.3%, 23.0%.

Biological parameters of hybrids fed under laboratory conditions are given in Table 3.6.6. As components for hybrids, among others, widely rayonised breeds Ipakchi 1 x Ipakchi 2, AGU-112 x UzNIISH 9 are used.

Table 3.6.6 Biological indicators of hybrids (2013-2015)

No. of items	Name of hybrids	Caterpillar life, %	Mean weight		% silk shell
			cocoon,g	shell, mg	
1	Ipacci 1 x Ipacci 2	95,3	1,85	453	24,5
2	Ipacci 2 x Ipacci 1.	93,1	1,86	461	24,8
3	Sword.1 x Sword.2	97,1	1,78	429	24,1
4	Sword.2 x Sword.1	95,6	1,95	438	23,7
5	SANIIISH 30 x C-5	96,9	1,98	451	22,8
6	C-5 x SANIISH 30	94,0	1,96	449	22,9
7	AGU-112 x UzNIISH 9	93,1	1,78	416	23,4
8	UzNIIISH 9 x AGU-112	98,0	18,6	430	23,1
9	C-13 x C-14	93,1	1,73	438	25,3
10	C-14 x C-13	72,7	1,76	434	24,7
11	Ipakci 2 x C-5	94,2	1,92	457	23,8
12	C-5 x Ipakci 2	95,5	1,97	464	23,6
13	Ipakci 1 x Atlas	95,6	1,76	425	24,1
14	Ipakci 2 x Atlas	96,2	1,88	46,8	24,9
15	Atlas x SUNYISH 30	97,3	1,94	465	24,0
16	C-12 x C-10	96,7	1,86	471	25,3
17	7 pk x C-10	95,3	1,92	443	24,0
18	9 pk x C-14	95,8	1,90	449	23,9
19	51.40 pk x C-5	96,4	1,84	437	23,8
20	AIC x C-12	96,0	1,69	396	23,5
21	22 pk x C-5	95,7	1,80	423	23,5
22	23 pk x C-10	94,5	1,86	439	23,8
23	PC x L-48	97,7	2,01	477	23,7
24	PC x L-51	95,8	2,03	478	23,5
25	Atlas x Asaka (k)	95,3	2,04	485	23,8
26	Margilan x Atlas (k)	96,7	1,97	462	23,5
27	Asaka x Marhamat (k)	93,6	1,94	466	24,0
28	Marhamat x Asaka (k)	93.8	2,01	463	23,0

Twenty-eight hybrids were created from different genotyped breeds. They were fed in laboratory conditions, on the same feed, under normal hydrothermal regime. Good results were obtained. Caterpillar viability 93.1-98.0%, cocoon weight 1.73-2.04 g, silkiness 22.8-25.3%. From 4-clonal hybrids two hybrids were zoned in Tashkent region. These hybrids as promising hybrids were tested in GSA.

Ipakchi 1 x Ipakchi 2 hybrids were released in Uzbekistan more than 20 years ago. The components of hybrids are Ipakchi 1, Ipakchi 2 breeds.

The breed Ipakchi 1 is bred by synthetic selection. It has the following indices: 98.6 per cent grena revival, 89.9 per cent caterpillar viability, 24-25 per cent silkiness, oval-round cocoon shape. Caterpillars go to curl in a friendly manner. Average yield in laboratory conditions 70-75 kg.(Figure 3.6.3).

Fig. 3.6.3 Cocoons of the Ipakchi breed 1

The breed Ipakchi 2 is bred by the method of synthetic selection, has the following indicators: caterpillar revival 97.6%, caterpillar viability 89.9%, silkiness 24-25.5%. Caterpillars go to curling in a friendly manner. The shape of the cocoon is with a slight interception. The average yield under production conditions when fed with varietal leaf is 7075 kg per one box of caterpillars (Figure 3.6.4).

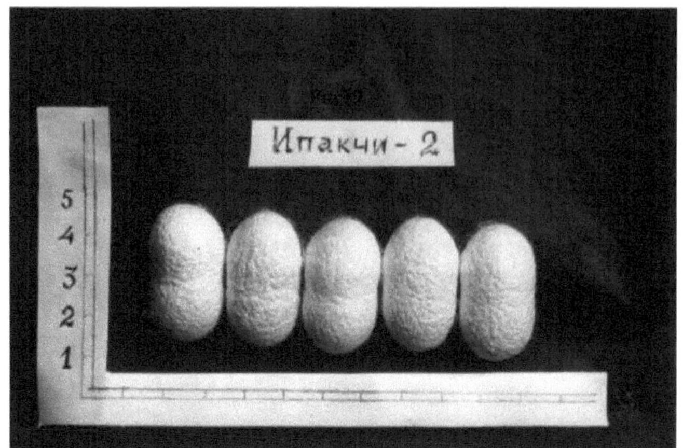

Fig. 3.6.4 Cocoons of the Ipakchi breed 2

Fig. 3.6.5 Cocoons of hybrid Ipakci 1 x Ipakci 2

Grena of hybrids Ipakci 1 x Ipakci 2, Ipakci 2 x Ipakci 1 are dark-coloured, homogeneous, contain more than 600-700 pieces of grena in one clutch.

Caterpillars of hybrids Ipakci 1 x Ipakci 2, Ipakci 2 x Ipakci 1

simultaneously go to curling, curling outwardly homogeneous cocoons.

Cocoons of Ipakci 1 x Ipakci 2 slightly oval in shape, hybrid Ipakci 2 x Ipakci 1 - with a slight intercept, uncoiling rate 89-90%, metric thread number over 3100-3150 units. Length of continuously unwinding cocoon from 950-1100 metres. Under production conditions, the yield is 60-65 kg, but, when feeding good varietal silkworms-more, the yield reaches an average of 70-75 kg. from one box of grena (Fig.3.6.5).

§ 3.7 Station tests of hybrids in conditions of the Republic of Karakalpakistan

In order to properly assess the quality of hybrids proposed for introduction, it is necessary to familiarise ourselves with the results of coconut harvesting in the

Republic of Karakalpakistan. Table 3.7.1 provides information on yields of mulberry silkworm hybrids currently reared in Karakalpakstan.

Table 3.7.1.

Information on harvesting of mulberry silkworm cocoons in 2015, 2016, 2017 by districts of the republic

Karakalpakistan

Neighbourhoods	Grena distribution, boxes			Cocoons harvesting plan, tonnes			Harvested cocoons, tonnes			Plan fulfilment, %			Yield of cocoons 1 cor.gr, kg		
	2015	2016	2017	2015	2016	2017	2015	2016	2017	2015	2016	2017	2015	2016	2017
Turtkul	3273	3273	3273	186,8	186,8	186,8	187,3	187,0	187,0	100,3	100,1	100,1	57,2	57,1	57,1
Berunium	2866	2866	2866	164,8	164,8	164,8	165,3	165,5	165,4	100,3	100,4	100,3	57,7	57,7	58,0
Ellikal'a	3013	3013	3013	172,8	172,8	172,8	173,5	173,5	173,5	100,4	100,4	100,4	57,6	57,6	57,6
Amu Darya	4462	4462	4462	253,3	253,3	253,3	253,9	253,9	253,9	100,2	100,2	100,2	56,9	56,9	57,0
Huzhaili	753	753	753	42	42	42	42,2	42,2	42,2	100,0	100,0	100,0	56,0	56,0	56,1
Shumanium	44	44	44	2,6	2,6	2,6	2,6	2,6	2,6	100,0	100,0	100,0	59,1	59,1	60,7
Canticle	27	27	27	1,6	1,6	1,6	1,6	1,6	1,6	100,0	100,0	100,0	59,3	59,3	59,3
Nukus	27	27	27	1,6	1,6	1,6	1,6	1,6	1,6	100,0	100,0	100,0	59,3	59,3	59,3
Kegaili	329	329	329	15,8	15,8	15,8	15,9	15,9	15,9	100,6	100,6	100,6	48,3	48,3	48,6
Chimbay	206	206	206	8,5	8,5	8,5	8,6	8,6	8,6	101,2	101,2	101,2	41,7	41,7	41,7
Total	15000	15000	15000	850,0	850,0	850,0	852,6	852,6	852,6	100,3	100,3	100,3	56,8	56,8	56,7

As shown in Table 3.7.1, the yield of mulberry silkworm hybrids was not high and was 56.8kg in 2015, 56.8kg in 2016 and 56.7kg per 1 box of grena in 2017. In some districts such as Kegaili it was 48.3-48.6kg and in Chimbai it was 41.7-4.2kg. As is known, the average cocoon yield in Uzbekistan in 2017 turned out to be 57.8 kg, the highest yield was observed in Surkhandarya region 58.1 kg. In order to increase cocoon yield in Karakalpakstan, it is necessary to introduce new hybrids of mulberry silkworm adapted to extreme economic conditions.

The hybrids are tested in small-scale production trials before being passed on to production.

Fig.3.7.1 Feeding of caterpillars of clonal-breeding hybrids with leaves of varietal silkworms

Fig. 3.7.2 Production of cocoons of clonal hybrids under farm conditions

Table 3.7.2 summarises the results of selection trials for all hybrids tested under production conditions. We are only interested in the hybrids involved in our study. These are hybrids with components: 51.40pc. C-5, A-153PC, C-5ngl, Line 22.

The results of station tests (Table 3.7.2) show that the percentage of viability of 15 hybrids is 82.7-95.7 among hybrids with high indicators there is a hybrid with parthenoclon 51.40, which viability is at the level of the control hybrid Tetra-3, silkiness 27.3%, in the control 21.9%. For this trait, all hybrids have higher indicators than in the control. The obtained materials are mathematically processed and reliable (Table 3.7.2).

If hybrids show good grena revival, caterpillar viability and technological performance, they are submitted for State trials.

All hybrids with parthenoclonians are characterised by good viability - 90.5-95.7% (95.7% in the control) and high silkiness - 23.3-27.8%. Low coefficients of variation of clonal-breed hybrids in caterpillar viability (2.29-2.68) and silkiness (1.29-3.35) attract attention. This once again emphasises the stabilising role of parthenoclones in hybridisation.

High indicators of the main productive characteristics of clonal-breed hybrids - viability and silkiness, with low coefficients of variation, make it possible to recommend these hybrids for industrial fattening in ecologically difficult conditions.

Table 3.7.2.

Station tests of hybrids in conditions of the Republic of Karakalpakstan and their evaluation (2015-2017)

No. No. of items	Name of hybrids	Caterpillar viability, %				Average cocoon weight				Average % silk sheath			
		M±t	A	c_v	Pd	M±t	5	c_v	Pd	M±t	5	c_v	Pd
1	5140pk x C-5	95,7±1,14	2,19	2,29	0	2,20±0,034	0,068	31,09	0,746	23,3±0,39	0,78	3,35	0,921
2	153 pk x C-5	90,5±1,21	2,43	2,68	0,998	2,06±0,031	0,003	3,06	0,999	24,5±0,158	0,316	1,29	0,992
3	AIC x C-14	96,2±1,21	2,43	2,52	0,649	2,34±0,08	0,16	6,84	0,881	23,4±0,34	0,68	2,90	0,932
4	AIC x Mechennaya	95,7±1,09	2,19	2,29	0	2,06±0,051	0,051	0,942	0,997	27,8±0,221	0,44	1,59	0,999
5	C-13 x C-14	83,5±2,55	5,10	6,11	0,999	2,18±0,034	0,068	3,12	0,881	23,3±0,075	0,151	0,648	0,932
6	C-14 x C-13	94,3±1,94	3,89	4,12	0,746	2,23±0,391	0,078	3,50	-	23,3±0,473	0,946	4,05	0,909
7	Sworded x	92,2±1,7	3,40	3,6	0,97	2,81±0,0	0,14	5,01	0,99	22,8±0,5	1,13	4,97	0,82

No.	Name												
	Sworn 2	0		8	5	75	1		9	68	7		3
8	Sworn 2 x Sworn 1	94,2±2,06	4,13	4,38	0,746	2,71±0,053	0,107	3,94	0,999	22,99±0,345	0,69	3,00	0,864
9	Nowruz 1	86,7±1,58	3,16	3,64	0,999	24,5±0,073	0,146	6,05	0,971	24,4±0,029	0,058	0,239	0,999
10	Nowruz 2	84,8±2,43	4,86	5,73	0,999	2,38±0,474	0,095	3,98	0,989	23,8±0,486	0,972	4,08	0,955
11	Ipacci 1 x Ipacci 2	82,7±3,28	6,56	7,93	0,999	2,46±0,099	0,199	8,10	0,993	23,3±0,078	0,156	0,667	0,947
12	Ipacci 2 x Ipacci 1.	83,0±0,97	1,94	2,34	0,999	2,57±0,036	0,073	2,84	0,999	23,9±0,099	0,199	0,83	0,981
13	Tash. 6 x Osh 2	86,7±1,21	2,43	2,80	0,999	2,45±0,056	0,112	4,56	0,995	24,5±0,131	0,262	1,06	0,992
14	Osh 2 x Tash. 6	88,5±0,97	1,94	2,19	0,999	2,40±0,073	0,146	6,07	0,971	24,6±0,104	0,209	0,85	0,993
15	Tetrahybrid3 (control)	95,7±0,121	0,243	0,25		2,23±0,017	0,034	1,52		21,9±0,758	1,52	6,90	

The data presented in Table 3.7.2 show that clonal hybrids performed well, even slightly outperforming the control and other hybrids. The clone creators are confident in breeding new interesting, productive hybrids.

As already noted, hybrid C-13 x C-14 and its reverse combination since 1993 zoned in Bukhara and Andijan regions, hybrid Mezhenna 1 x Mezhenna 2 and its reverse combination zoned since 1994 in Andijan and Bukhara regions, hybrid Ipakchi 1 x Ipakchi 2 and its reverse combination zoned since 1995 in Andijan region. Below we present a table with production trials of hybrids.

Table 3.7.3.

Results of production trials of hybrids for three years

No. of items	Name of areas	Name of hybrids	Grenade factory	Number of boxes sold	Average lesson, kg
1	2	3	4	5	6
1	Republic of Karakalpakstan	T-3 (con)	K.Kurgan	1786	45,1
		Sword 1 x Sword 2	Urgench	500	46,4
		Sword 2 x Sword 1	Urgench	500	52,8
		Nowruz-1	Amu Darya		
		Nowruz-2	neighbourhood	4	58,0
		Ipakci 1 x		4	62,0
		Ipakci 2		4	62,0

1	2	3	4	5	6
		Ipakci 2 x Ipakci 1		4	61,0
2	Navoi region.	T-3 (con)	Navoi	3745	44,7
		C-3 x C-14	Navoi	215	50,7
		C-14 x C-13	Navoi	930	50,2
3	Samarkand obl.	T-3 (con)	Tashkent	5250	
		Sword 1 x Sword 2	K.Kurgan	1600	
		Sword 2 x Sword 1	K.Kurgan	1465	
		Ipakci 1 x Ipakci2	Samarkand	205	
			Samarkand	163	
4	Tashkent region.	T-3 (con)	Shahrisabz	2000	40,0
		C-13 x C-14	Navoi	715	50,5
		C-14 x C-13	Navoi	76	50,0
1	2	3	4	5	6
5	Khorezm region.	T-3 (con)			
		Sword 1 x Sword 2	Urgench	17047	51,6
		Sword 2 x Sword 1	Urgench	1315	51,1
			Urgench	1299	51,3
6	Through the Fergana Valley	T-3 (con)			
		Sword 1 x Sword 2		71792	53,7
		Sword 2 x Sword 1		250	68,0
				250	63,2

Table 3.7.3 summarises the results of testing the hybrids under study in different regions of Uzbekistan with different ecological conditions. It is all the more interesting to note that the average yield of cocoons from 1 box of grass in Karakalpakistan, a region with difficult climatic conditions, was not worse than in such regions as Fergana or Tashkent. For example, the yield of cocoons of hybrids Mech.1 x Mech.2, Mech.2 x Mech.1 in Karakalpakstan was 46.4%, 52.8%, and in Samarkand region - 54.2%, 52.6%. Yield of hybrids Ipakchi 1 x Ipakchi 2, Ipakchi 2 x Ipakchi 1 in Karakalpakstan was 62.0% and 61.0%. And in Samarkand region 61.9% and 62.0%.

Thus, the yield of the studied hybrids in extreme environmental conditions was at the level of yield of the same hybrids in other regions of Uzbekistan. This gives grounds to recommend for industrial foddering of Karakalpakstan hybrids: Mechenna 1 x Mechenna 2, Mechenna 2 x Mechenna 1, Ipakchi 1 x Ipakchi 2, Ipakchi 2 x Ipakchi 1.

In production conditions of Karakalpakistan hybrids Ipakchi 1 x Ipakchi 2, Ipakchi 2 x

Ipakchi 1, Navruz 1, Navruz 2 gave good results - 61.0 and 62.0 kg of cocoons were obtained from 1 box (in control Tetra-3 50.9kg). yield largely depends on fodder. It is necessary to improve the fodder ration with varietal mulberry trees. At present, varietal trees account for only 5-6% of the country's fodder. The main fodders are local Hasak and free pollination hybrid.

§ 3.8 Promising breeds and hybrids of mulberry silkworms promising for introduction of silk production in Karakalpakstan

In recent years, special attention is paid to the quality of raw silk. The most realistic method of improving the technological properties of produced cocoons is the creation and introduction of hybrids of mulberry silkworms that meet the requirements of the processing industry.

In strange developed silk breeding gives primary knowledge to breeding and multiplication of breeds with better technological properties.

A number of scientists [124,c 31],[125,c 16-30],[92,c 24-45],[95,c 18],[14,c 25] of Uzbekistan have studied textile properties of cocoons and proposed for introduction several hybrids with good technological parameters.

Silk production in Karakalpakistan also needs mulberry silkworm hybrids with high technological parameters of cocoon thread. Therefore, we considered it possible to consider other hybrids with interesting properties.

Taking into account the requests of the industry for supply of breeds and hybrids with high yield of silk products and increased cocoon unwinding, silk scientists have created a number of breeds and promising hybrids of mulberry silkworm with high textile properties. These breeds are Ipakchi 3 and Line 22.

Cocoons of Ipakci 3-white, elongated with slight interception, fine-grained (Fig.3.8.1), grena revivability-96.98%, caterpillar viability-88.91%.

The breed was developed by synthetic selection by hybridisation of foreign breeds with local ones, it is distinguished by homogeneous development of caterpillars and friendly curling of cocoons.

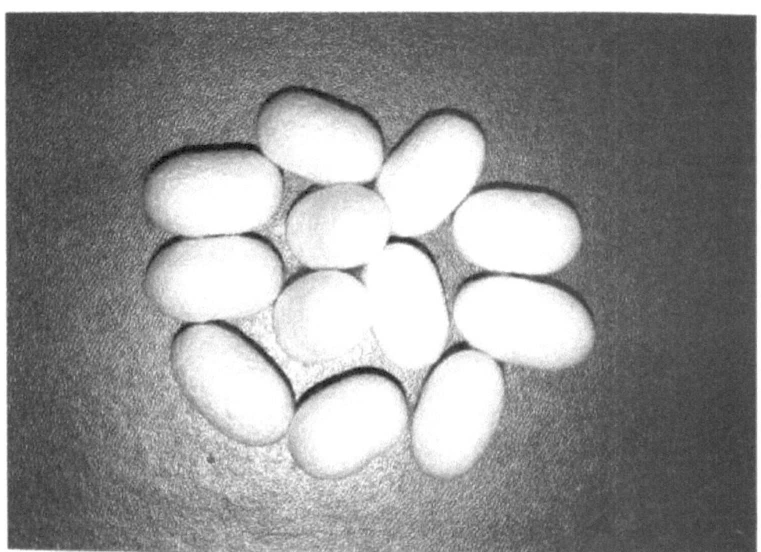

Fig. 3.8.1 Cocoons of the Ipakchi breed 3

Cocoons of Line 22 are white, oval-round, fine-grained (Figure 3.8.2), grena revival rate-96%, caterpillar viability-89%. The breed, bred by synthetic selection, is characterised by friendly curling of cocoons.

Fig. 3.8.2 Cocoons of the Line 22 breed

The hybrids between the L-22, Ipakchi 3 breeds were named Navruz-1, the reverse combination, Navruz-2.

The hybrid Navruz-1 was created at NIISH, in the laboratory of genetics and breeding of mulberry silkworm. It is a simple hybrid obtained from crossing Line 22 x Ipakchi

3. The grena is grey-petalled and the number of eggs in one gram is 1723 eggs. Percentage of grena revival-98%, number of caterpillars in 1 gram 2420 pieces. Caterpillars develop evenly, germinate on cocoons in a friendly manner. Cocoons are oval, viability of caterpillars-90.1%, yield of cocoons from 1 gram of caterpillars varies from 4.0 to 4.3 kg, silkiness of dry cocoons-56.8%, unwinding of the shell-88%. Raw silk yield-45.7%. Average length of cocoon thread - 1008 m, length of continuous unwinding thread-957 metres. Metric number of the yarn - 3500 unit (Figure 3.8.3).

Fig. 3.8.3 Cocoons of hybrid Navruz-1, zoned for production

The hybrid Navruz-2 was created at NIISH in the laboratory of genetics and breeding of mulberry silkworm. The simple hybrid is obtained from crossing of Ipakchi 3 and Line-22 breeds. The grena is of greyish colour, the number of eggs in one gram is 1708 pieces, on the first day of revival 97,5% of eggs come to life, the number of caterpillars in one gram is 2391 pieces, caterpillars develop smoothly, they sprout on cocoons in a friendly manner, the viability of caterpillars is 89,9%. Cocoons are oval with weak interception, fine-grained, white colour without shades (Figure 3.8.4), yield of cocoons from 1 gram of caterpillars - 4.73 kg, from 1 box of grens - 70-75 kg. Silkiness of dry cocoons - 55,7%, cocoon shell unwinding capacity - 88,4%, raw silk yield - 45,0 average length of cocoon thread -1058 metres, length of continuously unwinding thread - 948 metres.

Navruz-1 and Navruz-2 hybrids do not require any special zootechnical measures when feeding. The hybrids are characterised by high cocoon thread fineness - metric number Navruz-1 -3400 units, Navruz-2 -3300 units.

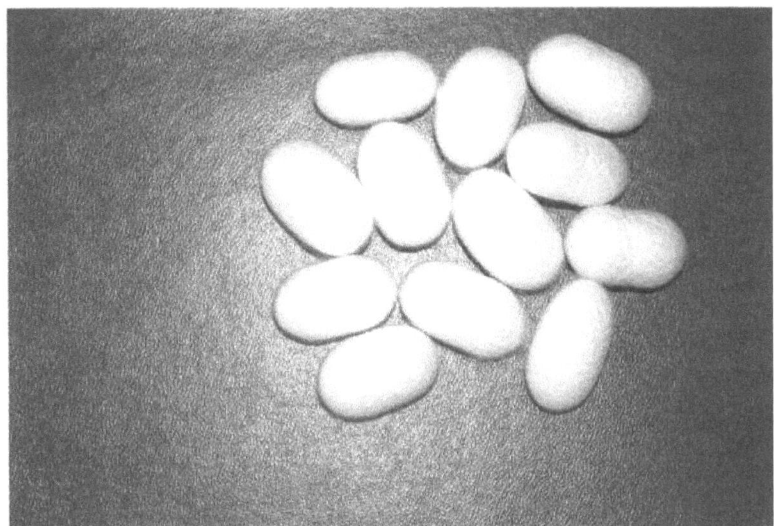

Fig. 3.8.4 Cocoons of hybrid Navruz-2, zoned for production

By the decision of the State Statistics Committee of Uzbekistan in 2013, Navruz-1 and Navruz-2 hybrids were recognised as promising, included in the State Register of Uzbekistan and recommended for introduction in Samarkand, Surkhandarya and Kashkadarya regions and others.

In small volumes, Navruz 1, Navruz 2 hybrids were foraged in Karakalpakistan. The results were encouraging (Table 3.7.2). High yields of Navruz 1, Navruz 2 hybrids - 58.0 and 62.0 kg in comparison with the control - 45.1 kg gives grounds to recommend them for foraging in industrial conditions of Karakalpakistan.

In 2012, some hybrids were reared in different regions of Uzbekistan. The results are summarised in Table 3.8.3.

Table 3.8.3.

Information on the performance of new silkworm hybrids under production conditions in 2012

Name of hybrids	Fed tu s.,cor.	Total cocoons collected, g	Average yield of cocoons,%	Variety cocoons, %	Raw silk yield, %	Length of coc.thread,m		Metric number of the thread
						general	NSD	
Andijan region								
Sword.1 x Sword.2	100	6,5	65,0	94,6	44,4	1203	416	3090
Sword.2 x Sword.1	100	5,6	56,0	93,4	43,9	1186	574	3072
Tetrahybrid 3	4-63	232,2	57,3	89,1	39,7	914	331	2810
Tashkent region								
51.40 PC x C-5	97	5,1	52,0	82,2	37,41	962	534	2990
Sword.1 x Sword.2	75	3,8	50,3	84,1	39,73	POZ	641	3058
Sword.2 x Sword.1	70	3,5	50,6	84,1	38,78	1092	633	3001
Tetrahybrid 3	21567	1069,1	49,6	79,5	33,17	812	316	2782
Bukhara region								
Sword.1 x Sword.2	54,1	33,3	61,6	93,8	40,2	986	491	-
Sword.2 x Sword.1	330	20,6	58,9	94,5	41,0	1010	534	-
Tetrahybrid 3	45928	2096,4	46,3	87,9	36,3	715	302	-
Jizzak oblast								
Sword.1 x Sword.2	316	15,4	48,6	94,9	43,41	1180	1054	2896
Sword.2 x Sword.1	249	13,9	55,7	95,5	41,85	1100	942	2819
C-13 x C-14	60	3,3	52,2	90,7	42,07	1250	1061	2851
C-14 x C-13	80	3,1	51,4	89,2	42,63	1226	1007	2885
Tetrahybrid 3	6985	352,1	50,4	81,1	32,43	734	371	2450
Fergana region								
51.40 PC x C-5	100	5,9	58,8	90,1	39,3	1041	723	2832
Sword.1 x Sword.2	560	38,3	68,4	92,2	42,3	1215	953	2962
Sword.2 x Sword.1	340	22,8	67,1	91,3	42,9	1268	1013	2976
C-13 x C-14	120	8,4	70,1	89,6	43,2	1193	834	2912

| C-14 x C-13 | 120 | <u>8,5</u> | <u>69,6</u> | <u>91,4</u> | 42,2 | 1210 | 996 | 2983 |
| Tetrahybrid 3 | 37959 | 2148,8 | 56,6 | 87,1 | 32,1 | 844 | 396 | 2843 |

Table 3.8.3 shows that hybrids created using sex-labelled breeds and parthenogenetic clones showed yields higher than the control (from 48.6 kg in Mechenna 1 x Mechenna 2 in Jizzak oblast to 70.1 kg in C-13 x C-14 in Fergana oblast, with yields in the control from 46.3 in Bukhara oblast to 57.3 kg in Andijan oblast). This indicates high viability of hybrids with genetic rearrangements in genomes under production conditions. So, the use of highly productive, highly viable, genetically modified breeds of mulberry silkworm on industrial fattening can bring economic benefits and significantly reduce operations on separation of material by sex. Hybrids Mezhdenaya 1 x Mezhdenaya 2, Mezhdenaya 2 x Mezhdenaya 1, 51.40pk x C-5, C-13 x C-14, C-14 x C-13 can also be recommended for silkworm breeding in Karakalpakistan.

§ 3.9 Discussion of the results obtained

One of the priority branches of agriculture in Uzbekistan - silk breeding is gradually being revived anew. The diversity of natural-ecological zones in Uzbekistan urgently dictates the need for intensive development of new methods of selection and reproduction of many species of plants and animals adapted to different environmental conditions. The creation of breeds and hybrids of mulberry silkworm in less favourable weather-climatic and fodder conditions during the fattening period is of decisive importance in this matter. In the historical Decree of the President of the Republic of Uzbekistan PP-512 "On measures on further reforming of silk industry of the Republic" and in the Directive documents of the Cabinet of Ministers of the Republic of Uzbekistan set the tasks of transition to the production of industrial hybrid grena of domestic breeds and full provision of the need of the industry in high quality, adapted to the specific hot climate of our region, grena. Undoubtedly, the Republic of Karakalpakistan, Khorezm oblast, northern districts of Bukhara oblast, Fergana valley, Tashkent, Syrdarya, Jizzak, Samarkand, Bukhara, Navoi, Surkhandarya, Kashkadarya oblasts differ significantly not only in weather and climate, but also in soil characteristics, water availability and other conditions. In this aspect, the work on creation and introduction of mulberry silkworm hybrids for extreme environmental conditions is relevant.

More than 50 new breeds, lines and their hybrids have been created in the Research Institute of Silk Breeding by scientists-breeders V.A.Strunnikov, L.M.Gulamova, N.V.Shurshikova, R.K.Kurbanov, N.T.Chernetsova, H.Zakirova, A.I.Islamov and others. More than 25 new productive varieties of mulberry trees have been created, which are zoned in the republic. Their authors are A.F.Didichenko, S.S.Zinkina, M.I.Grebinskaya, O.Pulatov, Y.Miralimov, K.Shkalikova, U.K.Kuchkarov. With the help of induced mutations, highly productive breeds were obtained: C-5, C-5proz, C-10, C-12, C-13, C-14, SANIIISH-30, AG, MG, C-5ngl, C-9ngl. On the basis of these breeds, many industrial hybrids were created, which were zoned in different years in production. These hybrids are:

1. Meechenaya-1 x Meechenaya-2 - zoned
2. Meechenaya-2 x Meechenaya-1 - zoned
3. C-13 x C-14 - zoned
4. C-14 x C-13   - zoned
5. SANIISH-30 x C-5 - zoned
6. C-5 x SANIIISH-30 - zoned
7. C-12 x C-10 - prospective
8. C-10 x C-12 - prospective
9. UzNIIISH-9 x AGU-112 - zoned
10. AGU-112 x UzNIIISH-9 - zoned
11. Navruz 1 - zoned
12. Nowruz 2 - zoned
13. Nowruz 3 is promising
14. Nowruz 4 is promising

These hybrids are highly silk-bearing, with good technological properties, they give good yield and quality silk thread. NIISH scientists have also developed methods for obtaining meiotic and ameiotic parthenoclones. One of these clones 51.40pc in combination with the sex-labelled breed C-5 was zoned for production. Such hybrids as SANIISH-30 x Sov-5, Sov-5 x SANIISH-30, Asaka x Markhamat, Markhamat x Asaka, Tetrahybrid 3, Tetrahybrid 4 have been zoned in Uzbekistan since the 1960s. Since 1990, all these hybrids were gradually replaced by new hybrids: Ipakchi-1 x Ipakchi-2, Ipakchi-2 x Ipakchi-1, Uzbekistan 5, Uzbekistan 6. Recently, the arsenal of hybrids was replenished with new hybrids Navruz-1, Navruz-2. SANIISH-9 x TashSHII-112 and its reverse combination were recommended for summer feeding. The created breeds and hybrids are fed mainly in temperate zones of the republic. There is no specific hybrid of mulberry silkworm, as well as silkworm varieties for ecologically not favoured zones, such as Karakalpakistan, Surkhandarya oblast.

The main objective of the research work is to select the most adapted hybrids for Karakalpakstan from the previously created hybrids. For this purpose, several hybrids are tested annually in farms in Karakalpakstan.

A total of 36 hybrid combinations from different breeds and lines of mulberry silkworm were analysed. Among them: Ipakchi 1 x Ipakchi 2, Ipakchi 2 x Ipakchi 1, Navruz-1, Navruz-2 and many others.

Breeding and pedigree work was carried out with the components of the above hybrids in order to improve their main productive indicators. Breeding and pedigree work with breeds C-5, C-10, C-12 was carried out at the family level. Breeds C-5, 7, 9, 10, 12, SANIIISH 30, Ipakchi 1, Ipakchi 2, Asaka, Markhamat, Atlas, Margilan Mechnaya 1, Mechnaya 2 were bred in 2 - 10 repetitions. The main objective of the work was to improve economically useful traits of breeds and to preserve their genetic features.

Tagged breeds: C-5, C-7, C-9, C-10, C-12, C-13, C-14, SANIIISH-30M and unlabelled SANIIISH-30, Asaka, Markhamat, Atlas, Markhamat, Ipakchi 1, Ipakchi 2, as well as promising parthenoclon females: A-153 pk, A-238 pk, APK, 9pk, and 9pk were maintained and multiplied.

Breed C-13 is characterised by grena animation 95.4%. Families with 96,8% grena revival were selected for fattening, families with 97,8% grena revival at S=1,20, caterpillar viability 81,6% at S=6,50, cocoon weight 1,28 g at S=0,01, shell weight 352 mg at S=6,0, silk content 27,6% at S=0,1% were left for breeding. The mass of breeding cocoons was 1.40 g, sheath mass 400 mg, silk content 28.5% at S=0.42.

Breed C-14 is characterised by 87,1% grena revival, families with 95,7% grena revival were selected for fattening, families with 97,0% grena revival at S=1,30, with caterpillar viability 94,2% at S=23,10, cocoon mass 1,24 g at S=0,02, shell mass 372 mg at S=2,0; silkiness 30,0% at S=0,3 were left from the breeding. The mass of breeding cocoons was 1.42 g at S=0.05, sheath mass 415 mg at S=30.0, silkiness 29.22% at S=1.12.

Breed Mechnaya-1. It is characterised by grena revival of 97.4%, families with grena revival of 98.0% were selected for fattening. Families with grena revival 98,2% at S=0,20, caterpillar viability 89,2% at S=1,28, cocoon weight 1,28 g at S=0,1 g, shell weight 297 mg at S=2,0, silkiness 23,2% at S=0,2 were left for breeding. Breeding cocoons weight 1.47 g at S=0.07, sheath weight 370 mg, at S=45.0, silk content 25.17% at S=1.97.

Breed Mechnaya-2. It is characterised by grena revival 98,4%, families with grena revival 98,5% were selected for fattening, families with grena revival 98,8% at S=0,4, caterpillar viability 87,8% at S=4,70, cocoon weight 1,33 g at S=0,02, shell weight 304 mg at S=5,0, silk content 22,9% at S=0,1. Breeding cocoons weight 1.50 g at S=0.05, silk sheath weight 375 mg, at S=15.0, silk content 25.0% at S=0.18.

The C-5 breed was reared piecemeal families were formed from clutches with 95.7% clutch revival, 72.6% caterpillar viability, cocoon weight 1.43 g, shell weight 366 mg, and 25.8% silkiness.

Breed C-10 was reared singly, families were formed from clutches with 96.4% clutch revival, 86.7% caterpillar viability, cocoon weight 1.35 g, shell weight 343 mg, and 25.6% silkiness.

The C-12 breed was reared piecemeal families were formed from clutches of grens, whose revival was 916%, caterpillar viability 80.4%, cocoon weight 1.36 g, shell weight 381 mg, and silkiness 29.9%.

The breed SANIISH 30 was reared in 7-10 replicates, the replicates were formed from clutches of grens with 97.8% revival, 88.3% viability, cocoon weight -1.35 g, shell weight - 298 mg, silkiness-22.4%.

The breed SANIISH 30 labelled at the caterpillar stage was reared in 3-6 replicates, the replicates were formed from a mixture of clutch clutches with clutch viability 96.8%, caterpillar viability 80.5%, cocoon weight 1.30 g. shell weight 277 mg, silkiness 21.3%.

The Asaka breed was reared in 6-8 replicates, the replicates were formed from a mixture of clutch grens, with 97.3% revival, 92.1% viability, 1.38 g cocoon weight, 289 mg shell weight, and 21.1% silkiness.

Marhamat breed was reared in 6-8 repetitions. Repeats were formed from clutches of grens, whose revival was 97.0%, caterpillar viability 83.4%, cocoon weight 1.25 g, shell weight 249 mg, silkiness 20.3%.

The breed Margilan was reared in 4 replicates, the replicates were formed from clutches of grens with 97.5% viviparity, 88.7% caterpillar viability, cocoon weight 1.51 g, shell weight 345 mg, silkiness 23.0%.

The Atlas breed was reared in 2-4 replicates, the replicates were formed from a mixture of clutches of grens with 96.5% revival, caterpillar viability 71.0%, cocoon weight 1.42 g, shell weight 333 mg, and silkiness 23.3%.

The breed Ipakchi 1 was reared in 4 replicates, the replicates were formed from a mixture of clutch grena with revival 93.0%, caterpillar viability 72.3%, cocoon weight 0.990 mg, shell weight 235 mg, and silkiness 23.7%.

The breed Ipakci 2 was reared in 4 replicates, the replicates were formed from a mixture of clutches with revival 95.2%, caterpillar viability 85.2%, cocoon weight 1.19 g, shell weight 346 mg, silkiness 21.4%.

The breed UzNIISH-9 was bred from the bivoltine breed SANIISH-9 and reared in 4-6 repetitions. During the years of breeding work, the viability of caterpillars was improved from 86.3% to 91.4%, average cocoon weight from 1.52g to 1.91g, silkiness of cocoons from 21.0% to 24.6%.

The breed AGU-112 was created from bivoltine breed TashSKHI-112 and reared in 4-6 repetitions. Selective breeding resulted in improvement of: caterpillar viability - from 89.4% to 91.8%, cocoon weight - from 1.52g to 1.88g, cocoons silkiness - from 20.4% to 24.4%.

All the above mentioned breeds, after carrying out selection and pedigree selection with them at all stages of development, served as components of different-type hybrids of mulberry silkworm, passed laboratory and small-scale production tests. On the basis of analysis of their productive characteristics, simple hybrids can be recommended for breeding in conditions of Karakalpakistan: UzNIISH 9 pk x AGU-112, Navruz 1, Navruz 2, Ipakchi 1 x Ipakchi 2, Ipakchi 2 x Ipakchi 1, hybrids with participation of genetically modified breeds: C-13 x C-14, C-14 x C-13, Sworded 1 x Sworded 2, Sworded 2 x Sworded 1, clonal-breed hybrids: APK x C-14, APK x MG, 9PK x Ya-120.

The hybrids UzNIISH-9 x AGU-112, AGU-112 x UzNIISH-9 were created specifically for growing under extreme conditions of late-spring, summer, summer-autumn fodder [52;p.3-19]. The yield of these hybrids in the summer season of 2015 was 46 kg per 1 box of caterpillars, against 44 kg in the control. The component breeds UzNIIISH-9, AGU-112 of these hybrids are bivoltine breeds, i.e. they have increased resistance to unfavourable environmental effects. Therefore, hybrids between them give high yields and can be used for foraging in ecologically difficult areas of Uzbekistan, including Karakalpakistan.

Hybrids Ipakchi 1 x Ipakchi 2, Ipakchi 2 x Ipakchi 1 are widely zoned throughout Uzbekistan. For more than 20 years, these hybrids show consistently high yields - 65-

70 kg per 1 korghrena, against 56.9 kg in the control. Component breeds Ipakchi 1, Ipakchi 2 of these hybrids, being products of synthetic breeding, carry parthenoclon genes in their genotypes, i.e. they have high viability. Therefore, hybrids between them are able to resist unfavourable climatic conditions and can be used for foraging in ecologically difficult zones of Uzbekistanisotan, including Karakalpakistan.

Navruz 1, Navruz 2 hybrids were created as hybrids with high technological characteristics of cocoon yarn. Metric yarn number of Navruz 1-3584 units, Navruz 2-3717 units. Component breeds Ipakchi 3, Line 22 of these hybrids have good technological qualities of silkworm and high viability of caterpillars. Therefore, in order to improve the quality of silk produced in Karakalpakistan, they can be used for foraging in the northern regions of Uzbekistan.

Hybrids C-13 x C-14, C-14 x C-13 have 100% purity of hybrid grena preparation and high viability of the C-13 component breed. C-14 of these hybrids are sex-determined by the colour of the grena, and thus can be divided into females and males at the egg stage. The yield of hybrids - 65-70 kg from 1 box of grena, against 56.9 kg in the control, is achieved due to the manifestation of high heterosis. Economic efficiency from the use of hybrids C-13 x C-14, C-14 x C-13 can be very high, because in the process of preparation of hybrid grena reduces the operation of separation of cocoons by sex and grena is prepared with 100% purity. In addition, these hybrids have high viability of caterpillars, so they can be used for foraging in ecologically difficult areas of Uzbekistan, including Karakalpakistan.

Hybrids Meechennaya 1 x Meechennaya 2, Meechennaya 2 x Meechennaya 1 are highly productive hybrids [114;p.3-20]. Component breeds Meechennaya 1, Meechennaya 2 are sex-determined by the colour of caterpillars, so they can be divided visually into females and males at the caterpillar stage. High yield of hybrids-60-65 kg of cocoons from 1 box of caterpillars is achieved due to 100% purity of hybrid grena preparation and manifestation of powerful heterosis. The combination of high productivity of hybrids with high viability of their caterpillars makes it possible to use these hybrids in many regions of Uzbekistan, including Karakalpakistan.

Clonal-breed hybrids AIC x C-14, AIC x MG, 9pc x Ya-120 are distinguished by purity of grena preparation, strong heterosis, high viability and friendly development of caterpillars, uniformity of cocoons in calibre, good yield - 60-65 kg of cocoons from 1 box of caterpillars, against 56.9 kg in the control. Component breeds C-14, MG are labelled by sex at the egg (C-14) and caterpillar (MG) stages, which means that they can be divided into females and males at early stages of mulberry silkworm development and be delivered to mulberry plants bypassing the complicated and inaccurate operation of cocoons separation by sex. The breed Ya-120 possesses very thin cocoon thread - 4506 units and in hybrids with clones shows high technological qualities of cocoon thread. Parthenogenetic clones AIC, 9pk are characterised by high combinative ability and good viability, are represented by one sex - female, so they do not need to be separated by sex. The advantages of clonal-breeding hybrids make them very attractive for breeding in all regions of Uzbekistan, including Karakalpakstan.

The hybrids listed above have never been bred in Karakalpakistan before. Currently produced on the territory of the autonomous republic mulberry silkworm hybrids have lost their relevance and need to be replaced. In this regard, we propose to carry out breed change of mulberry silkworm in Karakalpakistan, replacing the currently used hybrids with those proposed in this study.

To increase the yield of mulberry silkworm cocoons in Karakalpak Republic, it is necessary to improve the agrotechnical care of silkworm with the introduction of new zoned varieties: Tajik seedless, Mankent, Zimostoyky, Surkh-tut, Jar-Aryk 4, Jar-Aryk 6, Jar-Aryk 7, Jar-Aryk 8.

In mulberry trees, especially in ecologically unfavourable regions, there are mutations that ensure adaptation of plants to changed growing conditions. In Karakalpakistan, 65-70 years ago the Aral Sea was near Nukus, and now the sea has moved away from Nukus about 150-160 kilometres. During this time, under the influence of the external environment, there were changes (mutations) in the mulberry population. It is necessary to search for such "Natural mutations", allowing plants to adapt to the changed conditions. to reproduce and use them to breed new high-yielding varieties of mulberry.

CONCLUSION

1. It is established on the basis of study of literary sources, scientific reports and experience of producers that the main factors providing success of fattening in extreme ecological conditions are breeds and hybrids of mulberry silkworm adapted to less favourable breeding conditions.

2. Breeds for use as source material in selection of breeds and hybrids intended for breeding in extreme conditions were selected: C-5, C-5ngl, C-10, C-12, C-13, C-14, Asaka, Markhamat, Ipakchi 1, Ipakchi 2, Mechnaya 1, Mechnaya 2, parthenoclones by analysing the main productive properties of labelled and unlabelled breeds maintained in the living collection of mulberry silkworm.

3. Improved in terms of caterpillar viability, cocoon weight and silkiness, resistant to unfavourable environmental conditions, mulberry silkworm lines were created using C-5, C-10, C-12, C-13, C-14 breeds as source material. Sword 1, Sword 2, SANIISH 30, SANIISH 30mech, Asaka Markhamat, Margilan, Atlas, Ipakchi 1, Ipakchi 2, by methods of traditional selection and selection by motor activity.

4. 36 different-type hybrids of mulberry silkworm were tested on the main economic-valuable indicators in laboratory conditions of Nukus branch of TashGAU in order to determine the most productive and resistant to environmentally unfavourable conditions of maintenance. The investigated hybrids have the following indicators: 96-98% grena revival, caterpillar viability 89,90%, cocoon weight 1,60-2,0 g, shell weight 400-430 mg, silkiness 21,0-26,0%, weight of one dry cocoon 0,75-0,96 g, yield of raw silk
47.20%-48.0%, silk yield 49.18-50.20%, sheath unwindability 89-91.8%, metric thread number 2710-3134 kg, continuous unwind length 754-1022 m, production length 1200-1435 m.

5. Hybrids for successful use on foraging in extreme conditions of Karakalpakstan were determined: Ipakchi 1 x Ipakchi 2, Ipakchi 2 x Ipakchi 1, Navruz 1, Navruz 2. These hybrids produce the highest yield from 1 box of caterpillars: Ipakchi 1 x Ipakchi 2 - 61 kg, Ipakchi 2 x Ipakchi 1 - 62 kg, Navruz 1 - 62 kg, Navruz 2 - 63 kg in control 55 kg.

6. It was found that hybrids C-13 x C-14, C-14 x C-13, Sworded 1 x Sworded 2, Sworded 2 x Sworded 1, AIC x C-14, AIC x Sworded 1, created with the participation of parthenoclones and sex-labelled breeds, were sufficiently resistant to extreme weather and climate conditions due to the fact that they consisted of highly heterosis caterpillars of 100% hybrid origin.

7. It was found out that hybrids of mulberry silkworm, recommended for breeding in Karakalpakistan, showed the greatest resistance to extreme conditions in terms of caterpillar viability: AIC x C 14 - 96,2 %, AIC x Mechnaya 1-95,74 %, C-14 x C-13- 94,3 % against 93,7 in control.

Practical recommendations

1. It is recommended to use the following hybrids for industrial forage in extreme ecological conditions: Ipakchi 1 x Ipakchi 2, Ipakchi 2 x Ipakchi 1, Navruz 1, Navruz

2, which are characterised by high viability and yield.

2.   It is also recommended, subject to strict compliance with the optimal maintenance regime, to breed in production in disadvantaged climatic regions hybrids C-13 x C-14, C-14 x C-13, Mechenna 1 x Mechenna 2, Mechenna 2 x Mechenna 1, AK x C-14, APK x Mechenna 1, which are characterised by 100% purity of preparation and show heterosis power on viability.

3.   It is recommended that in order to maintain the whole number of caterpillars in the nursery and increase silk productivity, varietal silkworms should be used as fodder.

LIST OF REFERENCES

1. Decree of the President of the Republic of Uzbekistan PP-2856 "On measures to organise the activities of the Association Uzbekipaksanoat" from 29.03.2017. - C.1-15.

2. Uzbekistan President 2015 and 29 December 2015 "2016-2010 Kishlok Khuzhaligini kishloq Khujaligini yanada isloh kilish va rivozhlantirish yaora-tadbirlari tuFrisida" gi PK-2460-sonli karori, 5-7-6.

3. Uzbekiston Respubliki Vazirlar Mahkamasining "2017-2010 yillarda pillachilik tarmoFini complex rivozhlantirish chora-tadbirlari dasturi touFrisida" gi 2017 yil 11 Augustdagi No. 616-sonli karori, 5-6-6.

4. Nasirillaev U.N., Lezhenko S.S. Basic methodological provisions of breeding work with mulberry silkworm (Guiding document). //Tashkent 2002.-C.3-16.

5. Abbasov B.G. Selection-genetic parameters of economically useful traits of zoned breeds of mulberry silkworm Azad and Ganja 1. //Diss.kand.biol.nauk - Kirovobod, 1975. - C.81-91.

6. Azimov S.G. Inheritability of productive traits of chickens cross 288. //Leningrad-Pushchino - 1973.-S.23.

7. Azimov S.G. Inheritability of egg production and egg weight in hens cross 288. //Theses of reports of the X-republican scientific and production conference of scientists and producers on animal husbandry. - Tashkent, 1975. -C.18.

8. Azimov S.G. Breeding work in poultry farming. //Publishers' Publishing House "Uzbekistan". - Tashkent, 1980.-S.9-61.

9. Azimov S.G. Raising high-yielding chickens in Uzbekistan. // Publication of UzNIINTI. - Tashkent. 1973. -C-1-3.

10. Azimov S.G. Results of crossing lines of parental forms of egg crosses "Zarya-17". // Theses of reports at the VI Congress
Uzbek Society of Geneticists and Breeders. - Tashkent, 1992. - C.17.

11. Azimov S.G. Selection of hens of egg lines to increase egg production under hot climate conditions. // All-Union conference on animal breeding. - Izd-vo TSKhA. - Moscow, 1979.- P.32.

12. Azimov S.G.. Rizaev A.A. Increase of weight of hens of selection of UzNIIZh at crossing with roosters of large-egg cross "Zarya 17". //Proceedings of the Uzbek Research Institute of Animal Husbandry "Ways to increase the productivity of livestock and poultry farming". - Tashkent, 1993.- P.125-131.

13. Azimov S.G. Scientific bases and methods of increasing productivity of hens of egg lines in conditions of hot climate. //Author abstract of disc.dok.s/khnauk.- Tashkent, 1994.-C.13-21.

14. Azimov S.G., Rizaev A.A., Azimov D.S., Nasibullina F.Sh., Arindilanov T.R. UzCHITI selection tovuklari tukhumlarining aegiluvchan deformation. //Uzbeki ston iktisodiy islohotlar davrida chorvachilikni rivozhlantirish ilmiy va amaliy asoslari. - Tashkent, 1996. - 102-105-6.

15. Azizov B.S. Tut ipak kurtining yirik pillali zotlari ishtirokidagi sanoatbop duragailari. Uzbekiston ipakchiligi rivozhlantirishining ilmiy asoslari "Fan" nashriyoti,

2001. -34-б.

16. Akizhikov Ya.S. Effect of selection of cocoon batches on silkiness at the reception at the drainage enterprises. //Silk. 1994.-№1-2.-C.18- 19.

17. Akizhikov Ya.S. New methods of increasing silkworm breeding batches of cocoons on grenzavods. //Intensive methods of cultivation of silkworm and mulberry silkworm. Collection of scientific works of Tashkent Agricultural Institute - Tashkent, 1990. - C.66-68.

18 Alexandrov M.V. Significance of air heat content in silkworm ecology. //Silk.-Tashkent. 1964.-№3.-C.32.

19 Alexandrov M.V. On the heat factor in the ecology of silkworms and other poikilothermic organisms. //Silk.- Tashkent, 1965.- №3.-S.13-17.

20 Aliev A.G. Significance of provocative selection in increasing resistance of mulberry silkworm to jaundice. //Шелк.-Ташкент.- 1971.-№1.-C.26-28.

21 . 14. Alimova H.A., Burnashev I.Z., Gulamov A.E., Saidova R.A. Analysis of changes in linear density along the length of cocoon yarns of modern hybrids. //J.Technology of textile industry. - Tashkent. №6, 2000. -C.25.

22 . 15. Alimova H.A. State and development of nanotechnology in the textile industry. //The importance of science integration and solution of actual problems in the organisation of production in the enterprises of textile industry. -Margilan. 2017. - C.176.

23 Astaurov B.L. Artificial parthenogenesis in mulberry silkworm: experimental studies. //Moscow-Leningrad. Izd. of the Academy of Sciences of the USSR.-Moscow, 1940.-C.136-140.

24 Astaurov B.L. Testing of breeds and hybrids of the first generation in mulberry silkworm. //Москва-Ташкент, 1933.-C.3-21.

25 Astaurov B.L. Heredity and development. Selected works. //Moscow. Izdvo "Nauka". 1974.-C.122.

26 Astaurov B.L. Experiments on experimental androgenesis and gynogenesis in the mulberry silkworm. //Biological Journal. Moscow. 1937.-T.6.-№1.-C.77-81.

27 Astaurov B.L. The problem of sex regulation. //Publishers of "Science and Man".-Moscow, 1965.-#2.-P.344-367.

28 Astaurov B.L. Artificial parthenogenesis, experimental polyploidy and sex in bisexual animals VKN: Actual issues of modern genetics: // Izd. of MSU. 1996.-C.130-141.

29 Astaurov B.L. Selection for the ability to thermal parthenogenesis and obtaining improved by this trait parthenoclones of silkworm. //Genetics 1973, No.9. -C.9.

30 Astaurov B.L. Cytogenetics of mulberry silkworm development and its experimental control. -Nauka, 1968. -C.3-21.

31 Astaurov B.L. New data on artificial parthenogenesis in the USSR mulberry silkworm. //1936. №7. -C.277-280.

32 Afanasyev B.A., Pereldin N.S. Cage fur farming. //M. "Kolos". 1966.-C.35-42.

33 .Akhmedov N.A. Ipak kurti kharorat va khavo ekologik khabarnoma - ecological

bulletin 1999. -C.42.

34 Akhmedov N.A. Ipak kurti uruFini bahorgi zhonlantirish davrida embryo rivozhlanishini vaktincha tukhtatish muddatini kurning zhonlanishiga tasiri. //Ipak, 1998. №2. -C.11-12.

35 Ashirov M.I. Scientific bases and practical methods of improvement of pedigree and productive qualities of black-sheep cattle in conditions of hot climate. //Autoref.dis.dokt.s/kh.nauk. Tashkent, 1994. - C.10-15.

36 Braslavsky M.E., Golovko V.O., Yu ta in C.D. Viso-koproduktivsh shovkovychnogo shovkovypryad. // Agrarnaya nauka - virob-novitsvu. -Kshv, 2001. - C.19.

37 Braslavsky M.E., Stotsky M.I., Zhuravl V.B. Novi pbridi shovkovychnogo shovkopryad. // Agrarnaya nauka - virobnovitsvu. -Kshv, 2002. 22 -C.24.

38 Belov P.F. On prospects of selection of mulberry silkworm for resistance to nuclear polyhedrosis. //Proceedings of the Georgian Agricultural Institute. Tbilisi. 1972. Vyp.84.-S.223-228.

39 Belyaev D.K. Biological aspects of animal domestication. //Materials of the All-Union meeting on genetics and selection of new breeds of farm animals.-Almaata. 1970.-C.30-44.

40 .Berg R.L. Standardising selection in the evolution of the flower "Botanical" journal, 1956. vol.41, no.3. -C.124.

41 .Berg R.L. Further studies on stabilising selection in the evolution of the flower. "Biology" journal, 1958. vol.43, no.1. -C.-63.

42 Valiullina M.H. Some issues of the methodology of mulberry silkworm breeding. //Proceedings of SANIISH, "Measures to increase the productivity of mulberry silkworm". Tashkent, 1970. Publishing house "Fan", vol. VI. - C. 102-111.

43 Verbitskaya G.A. Breeding of jaundice-resistant lines of mulberry silkworm by heat treatment. // Silk. Tashkent, 1971. - №1-3. - C. 9.

44 Verbitskaya G.A. Some factors of increasing resistance of mulberry silkworm to jaundice. //Autoref. dissertation of candidate of s.h. sciences. Tashkent, 1972. - C.22.

45 .Terskaya V.N. Cytology of mulberry silkworm (BOMBYX MORI) matings during artificial parthenogenesis, Ontogenesis, 1979, Vol.10, No.3.

46 Voskonyan V.B., Galstyan G.A. Heritability and repeatability of the main selectable traits of Caucasian brown cattle. //Materials of the conference on genetics and selection of agricultural plants and animals of the Transcaucasian republics. - Publishing house "Elm". Baku. 1975. - C.184-185.

47 Gatin F.G., Ogurtsov K.S., Asamova N.N. New breeding variety of silkworm SANIISh "Mehnat" Tashkent, 1986. -C.27-29.

48 Golovko V.A., Krichenko I.A. Infectious diseases of mulberry silkworm in silk-producing regions of the world. Materials of scientific-practical conference "Problem questions of silk breeding development", Kharkov, 1993. -C.121-125.

49 Grebinskaya M.I., Gulamova L.M., Yakubov A.B. On the possibility of autumn-winter experimental foraging. //Ref.sb. "Silk" №1, 1972. - C-5.

50 Gulamova L.M. Breeding of heterosis forms of mulberry silkworm by parthenogenesis. Materials of II All-Union seminar-consultation on genetics and selection of silkworm and mulberry. Tashkent, 1981.-C.22-24.

51 Gumbatov I.M. Effective multiplicity of feeding of mulberry silkworm caterpillars. //Silk - Tashkent, 1983. №6.-C.12-14.

52 Gurova R   .A. Possibility of keeping a number of generations of androgenetic clones of mulberry silkworm. Journal "Silk", 1969. №2. -C.26-28.

53 .Gurova R   .A. Opportunities of reuse
Androgenetic males of the mulberry silkworm. // "Silk", 1969. №3.-C.25-27.

54 Gurova R.A. Obtaining androgenetic clones of mulberry silkworm - Abstract of the candidate's thesis - 1969.-C.3-27

55 Gurova R.A. Influence of breed composition of females on the success of androgenesis in mulberry silkworm. //"Silk", Tashkent, 1969. №4. - C.23-24.

56 Daniyarov U.T. Selection and improvement of breeds and hybrids of mulberry silkworm for repeated fattening. Abstract of candidate dissertation. 2010.- C.3-19.

57 Danshina E.V. Development of priiohpv forecasting I onmuMi3a^i life expectancy of productive onion-crylich comas on the application of shovkovichnogo I unpaired shovkopryadiv.: Author's thesis ...kand.s- x.nauk.-Kharkiv, 2002.- C-9-13.

58 .Dzhuraev D. Development of ways to increase productivity of fattening and quality of cocoons with application of different ingredients. Автореферат-Ташкент, 1994.-C.3-30.

59 .Ezhkov B.A., Ezhkova L.V. Intermediate feeding of mulberry silkworm. // Silk. Tashkent, 1961. №1. - C.11.

60 .Zworykina V.V. Stage sensitivity of mulberry silkworm to temperature factor. - Proceedings of SANISH, vol.1. - Tashkent. 1956. - C. 23-28.

61 Ilyina E.D., Kuznetsov G.A. Fundamentals of genetics and selection of fur-bearing animals. M. "Kolos", 1969.-C.43-50.

62 Kashkarova L.F., Yakubov A.B., Larkina E.A. Breeds of mulberry silkworm. //Tashkent, 2008.-C.3-100.

63 Kashkarova L.F., Umarov Sh.R. Diseases of mulberry silkworm diagnosis and prevention. //Tashkent, 2008.

64 Kenjaev B., Nasirillaev U.N. Interaction of genotype with environment in mulberry silkworm. - Report. 1 Inheritability of traits of mulberry silkworm productivity in different natural-climatic zones //Proceedings of SANIISH. Tashkent, 1976. issue 10. - C. 83-89.

65 Kerimova I.O. Efficiency of chemical mutagens in mulberry silkworm. Avtoref. dissertation of candidate of biological sciences. Tashkent, 1972.-C.1-25.

66 Kichinov T.J., Ashirov M.I. Growth and development of heifers of different genotypes. // Agriculture of Uzbekistan. Tashkent, 1998. №2. - C. 10.

67 .Kovalev P.A., Shevelova A.A. Butterfly emergence and its causes. In the book "Grainage and selection of mulberry silkworm". Izdvo "Uchitel" 1966. -C.-59.

68 Klimenko V.V., Lysenko N.G., Haoyuan Liang. Parthenogenetic cloning in

genetics and breeding of mulberry silkworm. //Ж. Rozvedennya I genetika tvaryn. Kharkov 2013. №47. -С.40-55.

69 Klimenko V.V., Zabelina V.B., Lysenko N.G. Intraclonal variability of mulberry silkworm / / / / Materials of the international conference devoted to the 75th anniversary of the birth of Academician Y.P.Altukhov. - Moscow 2011. - С.161-162.

70 Kuchkarov U.K., Asamova M. New highly productive variety of silkworm SANIISH-36. // Silk. Tashkent, 1984. №5. - С. 33-35.

71 Kuchkarov U.K., Gatin F.G., Khalmatov D.I., Akhmedova M. New highly productive variety of silkworm Jar-Aryk 2. // Silk. Tashkent, 1994. №3-4. - С.18.

72 Kuchkorov U., Yakubov A.B., Kholmatov D Tutsorlarni yoshartirish. //Izhtimoiy iktisodiy farmer journal.2012.1-2 sleep, 26-b.

73 Kuch "orov U., Kholmatov D., Akhmedova M. Ishlab chikarishga zhoriy etilgan yangi duragay tutlar va istikbolli tut navlari. //Uzbekiston ipakchiligi rivozhlantirishning ilmiy asoslari. Fan, 2011, 5-b.

74 Kuchkorov U., Ogurtsov K. "Sho-tut va Balkhi-tut". // "Mehnat" nashriyoti. Toshkent, 1989. 10-11-б.

75 Landau S.M., Dobrovolskaya G.N., Kireeva V.I. Study of the possibility of obtaining mulberry silkworm lines resistant to activation of latent nuclear polyhedrosis virus. // Molecular Biology. Moscow, 1979. №22. - С. 86-99.

76 .Larkina E.A., Yakubov A.B., Daniyarov U.T. Results of the study of the genetic nature of the motor activity of mulberry silkworm   // "Uzbek Biological Journal" 2012.
№6. -С.45.

77 Larkina E.A., Yakubov A.B. Combinational value of inbred lines of mulberry silkworm. // Uzbek Biological Journal №1. Fan 2011. - С.50.

78 .Larkina E.A., Yakubov A.B. Stability of indicators of breeds of the world collection of mulberry silkworm UzNIISH as a result of closely related breeding. //Ж. "Agroilm", №1, 2012. -С.45.

79 .Larkina E.A., Yakubov A.B., Nodiralieva N., Daniyarov U. Maintenance of sex-labelled breeds at the grena and caterpillar stages.
//J.Achievements of science and technology APK. Moscow "Kolos" №2. 2002. - С-35.

80 .Larkina E.A., Yakubov A.B., Daniyarov U.T. Catalogue Genetic fund of the world collection of mulberry silkworm of Uzbekistan. Tashkent, 2012.-С.4-66.

81 Larkina E.A., Nodiralieva N. Features of work with mulberry silkworm breeds labelled by sex. //Herald of Agrarian Science of Uzbekistan, No.2 (16), 2004 -С.66-68

82 Larkina E.A., Yakubov A.B. Advantages of using parthenoclones of mulberry silkworm in hybridisation. //Ipakchilik sohdsining dolzarb muammolari va ularni yangi tekhnologilarga asoslangan ilmiy echimlari. Materials of the republican scientific conference. -Tashkent 2012. -С.30.

83 M.E. Lobashov. What is genetics - Leningrad. - Izdvo "Znanie" 1969. -С.28.

84 Madaminov K.M. Average weight of mulberry silkworm eggs and its variations depending on the zone of grena preparation. // Silk. Tashkent, 1978. №6.-С. 9.

85 Madrakhimov F. Development of selection methods for resistance to mulberry silkworm jaundice virus. //Autoreferat Dis.kand.s/kh.nauk, 1968. -C.20.

86 Mametkuliev B. Application of the method of early prediction of mulberry silkworm variability in breeding work. //Some issues of silkworm breeding development in Turkmenistan - Ashgabat Ilm, 1975. - C.50.

87 Mametkuliev B., Martynova E.D. Prospective hybrids of silkworm. //Agriculture of Turkmenistan - Ashgabat, 1979. №8.-C.33-34.

88 Mamadaliev F.H. Scientific bases of increasing productive and breeding qualities of downy goats of Uzbekistan. // Abstract of doctoral dissertation, Tashkent, 1991. - C.28-30.

89 Navruzov S.N. On cocoon production in Uzbekistan. // Silk. Tashkent, 1991. №5.-C. 3.

90 Navruzov S.N. Labour victory of silkworm breeders of Uzbekistan. // Silk. Tashkent, 1993. №3-4.-C.3.

91 Navruzov S.N., Nasirillaev U.N. Ipak kurti she kapalaklarini tana ulchovlari buiicha tanlashning serpushtlikka ta'asiri Uzbekistan Republicsida ipakchilik makhsulotlariri sifatini oshirish yullari. Ilmiy conference materials - Tashkent, 1997. - C.32.

92 Nasirillaev U.N. Study of the quality of breeding grena at breeding stations and grenzavods. //Journal "Silk" № 4, 1967. -C.7.

93 Nasirillaev U.N. Theory and practice of mass selection in mulberry silkworm. //Author abstract for the degree of Doctor of Agricultural Sciences, 1978.-C.1-49.

94 Nasirillaev U.N., Lezhenko S.S., Dvoinikova T.N., Mustafaeva G.Y., Azizov B.S. Tut ipak kurtining tirik pillalari zotlari ishtirokida sanoatbop duragailari. Uzbekiston ipakchiligi rivozhlanishning ilmiy asoslari. Fan Publishing House 2001, 34-b.

95 Nasirillaev U.N., Lezhenko S.S., Umarov S.R., Azizov B.S. Yoz-kuz mavsumida eng yukori makhsuldorlik khusususiyatiga ega bulgan duragailarni sinash va tanlab olish. //Ipak ilmiy tehnik journal. Tashkent. 2002. №2. -C.13.

96 Nasirillaev B.U. Genetic bases of selection on morphological traits closely correlating with technological properties of cocoons of mulberry silkworm Bombyx mori L. -Tashkent, 2016.-C.24-45.

97 Nasirillaev U.N. Genetic bases of mulberry silkworm selection. - Tashkent, "Fan" Publishing House 1985. - C. 42-51.

98 Pashkina T.A. Inheritability and genetic interrelation of cocoon sheath unwinding with breeding traits in mulberry silkworms: Author's thesis of Candidate of Biological Sciences. Tashkent, 1987.-C.9-20.

99 .PashkinaT   .A. Methodological guidelines for determining the inheritability and correlation coefficients of cocoon sheath unwinding and use of these parameters in breeding. Tashkent, 1986.-C.18.

100 . Poyarkov E.F. About the existence of seasonal phases in mulberry silkworms Sb "New in biology of silkworms", Moscow, 1959, - P.54-63.

101 . Priyezhev G.V., Shadybekova D. Trihybrids of mulberry silkworm. // Scientific

bases of silkworm breeding development in Uzbekistan. Tashkent, 1978. Issue. 12. - C. 59.

102. Safonova A.M. Some causes of defectiveness of pupae and butterflies of mulberry silkworm and ways of their elimination. // "Silk", Tashkent. 1972, №1. - C.10-11.

103. Safonova A.M. Dependence of cocoon sheath properties on the position of caterpillars during curling. //"Silk", Tashkent. 1978. №2. - C.13-14.

104. Safonova A.M., Akizhikov Y., Dehkanov M. Influence of selection of cocoon batches on breeding and silkiness on productivity of industrial hybrids. // "Silk" №1, 1981. -C.17.